Lecture Notes in Computational Vision and Biomechanics

Editors

João Manuel R.S. Tavares
R.M. Natal Jorge
Address:
Faculdade de Engenharia da Universidade do Porto
Rua Dr. Roberto Frias, s/n
4200-465 Porto
Portugal
{tavares, rnatal}@fe.up.pt

Editorial Advisory Board

For further volumes:
http://www.springer.com/series/8910

Lecture Notes in Computational Vision and Biomechanics

Volume 4

This book is the first to be published under the Book Series "Lecture Notes in Computational Vision and Biomechanics (LNCV&B)".

The research related to the analysis of living structures (Biomechanics) has been a source of recent research in several distinct areas of science, for example, Mathematics, Mechanical Engineering, Physics, Informatics, Medicine and Sport. However, for its successful achievement, numerous research topics should be considered, such as image processing and analysis, geometric and numerical modelling, biomechanics, experimental analysis, mechanobiology and enhanced visualization, and their application to real cases must be developed and more investigation is needed. Additionally, enhanced hardware solutions and less invasive devices are demanded.

On the other hand, Image Analysis (Computational Vision) is used for the extraction of high level information from static images or dynamic image sequences. Examples of applications involving image analysis can be the study of motion of structures from image sequences, shape reconstruction from images and medical diagnosis. As a multidisciplinary area, Computational Vision considers techniques and methods from other disciplines, such as Artificial Intelligence, Signal Processing, Mathematics, Physics and Informatics. Despite the many research projects in this area, more robust and efficient methods of Computational Imaging are still demanded in many application domains in Medicine, and their validation in real scenarios is matter of urgency.

These two important and predominant branches of Science are increasingly considered to be strongly connected and related. Hence, the main goal of the LNCV&B book series consists of the provision of a comprehensive forum for discussion on the current state-of-the-art in these fields by emphasizing their connection. The book series covers (but is not limited to):

- Applications of Computational Vision and Biomechanics
- Biometrics and Biomedical Pattern Analysis
- Cellular Imaging and Cellular Mechanics
- Clinical Biomechanics
- Computational Bioimaging and Visualization
- Computational Biology in Biomedical Imaging
- Development of Biomechanical Devices
- Device and Technique Development for Biomedical Imaging
- Digital Geometry Algorithms for Computational Vision and Visualization
- Experimental Biomechanics
- Gait & Posture Mechanics
- Multiscale Analysis in Biomechanics
- Neuromuscular Biomechanics
- Numerical Methods for Living Tissues
- Numerical Simulation
- Software Development on Computational Vision and Biomechanics
- Grid and High Performance Computing for Computational Vision and Biomechanics
- Image-based Geometric Modeling and Mesh Generation
- Image Processing and Analysis
- Image Processing and Visualization in Biofluids
- Image Understanding
- Material Models
- Mechanobiology
- Medical Image Analysis
- Molecular Mechanics
- Multi-Modal Image Systems
- Multiscale Biosensors in Biomedical Imaging
- Multiscale Devices and Biomems for Biomedical Imaging
- Musculoskeletal Biomechanics
- Sport Biomechanics
- Virtual Reality in Biomechanics
- Vision Systems

Ugo Andreaus · Daniela Iacoviello
Editors

Biomedical Imaging and Computational Modeling in Biomechanics

Springer

Editors
Ugo Andreaus
Department of Structural and Geotechnical Engineering
Sapienza University of Rome
Rome, Italy

Daniela Iacoviello
Department of Computer, Control and Management Engineering Antonio Ruberti
Sapienza University of Rome
Rome, Italy

ISSN 2212-9391 ISSN 2212-9413 (electronic)
ISBN 978-94-017-7971-5 ISBN 978-94-007-4270-3 (eBook)
DOI 10.1007/978-94-007-4270-3
Springer Dordrecht Heidelberg New York London

Printed on acid-free paper

Springer is part of Springer Science+Business Media (www.springer.com)

Foreword

The contributions cover a wide field of scientific and cultural topics, ranging from remodeling of bone tissue under the mechanical stimulus up to optimizing the performance of sports equipment, through the patient-specific modeling in orthopedics, micro-tomography and its application in oral and implant research, computational modelling in the field of hip prostheses, image based model development and analysis of the human knee joint, kinematics of the hip joint, micro-scale analysis of compositional and mechanical properties of dentin, automated techniques for cervical cell image analysis, and biomedical imaging and computational modelling in cardiovascular disease.

The aim is to provide a comprehensive forum for discussion on the current state-of-the-art in the fields of biomedical image analysis and biomechanics, with articles covering theory, methodology, practical applications and new trends on the subjects of interest, along with multidisciplinary applications. The main topics are the following: Image Processing and Analysis; Medical Image Analysis; Computational Bioimaging and Visualization; Experimental Biomechanics and Numerical Simulation; Mechanics of hard and soft tissues; Biofluid mechanics; Mechanics of prostheses and implant-tissue interfaces; Macro/microscopic imaging for biomechanical investigations; Imaging and measurements in biomedical engineering; Applications of Computational Vision and Biomechanics; Development of Biomechanics Devices.

The audience for this book will be the researchers, Ph.D. students, and graduate students with multidisciplinary interests related to image analysis and understanding, medical imaging, biomechanics, simulation and modelling, experimental analysis.

Ugo Andreaus
Daniela Iacoviello

Contents

A Review of Automated Techniques for Cervical Cell Image Analysis and Classification

Marina E. Plissiti and Christophoros Nikou

Abstract Cervical smear screening is the most popular method used for the detection of cervical cancer in its early stages. The most eminent screening test is the Pap smear, which is based on the staining of cervical cells, using the technique that was first introduced by George Papanicolaou (Science 1942). With this screening technique, precancerous conditions and abnormal changes in cells that may develop into cancer are recognized. The widespread use of this test in developed countries has significantly reduced the incidence and mortality of invasive cervical cancer. In the last years, many methods have been appeared in the literature, which aim at the automated determination of the cytoplasm and the nucleus in these images. In this context, sophisticated image processing techniques and feature extraction and classification methods have been developed by several researchers, in order to derive useful conclusions for the characterization of the contents of the Pap smear images. In this work, an overview of the published techniques related to cervical smear screening is presented, in order to provide an integrated essay of the state of the art methods in the specific scientific field. Special focus has been paid on two main concepts with great research interest: the cell image segmentation and the classification techniques proposed for the characterization Pap smear images.

Keywords Pap smear images • Cell nuclei segmentation • Abnormal cell feature extraction

1 Introduction

The Pap smear is extensively used in gynaecology as a screening test for the detection of premalignant and malignant processes in the cervix. It consists of a sample of cells taken from the cervix with a special device, which are stained using

M.E. Plissiti • C. Nikou (✉)
Department of Computer Science, University of Ioannina, Ioannina 45110, Greece
e-mail: marina@cs.uoi.gr; cnikou@cs.uoi.gr

D. Iacoviello and U. Andreaus (eds.), *Biomedical Imaging and Computational Modeling in Biomechanics*, Lecture Notes in Computational Vision and Biomechanics 4,
DOI 10.1007/978-94-007-4270-3_1, © Springer Science+Business Media Dordrecht 2013

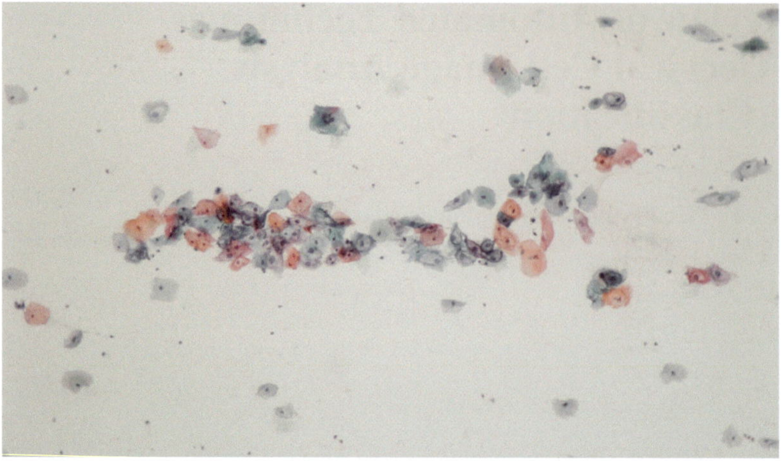

Fig. 1 A conventional Pap smear image

the Papanicolaou technique and smeared onto a glass slide. This slide is then examined by an expert cytologist under a microscope, in order to identify abnormalities in the structure and morphology of cell nuclei. Through this examination precancerous conditions and abnormal changes that may develop into cancer are recognized and as a consequence they are early treated preventing the development of cervical cancer.

Conventional Pap smears (Fig. 1) are used for over 60 years for screening of the cervical cancer and they exhibit certain characteristics, such as variances in illumination and dye concentration of the cells due to the staining procedure. Also, there are numerous variables, such as air-drying, excessive blood, mucus, bacteria or inflammation which make the recognition of the suspicious cells a difficult task. In terms of screening tests, liquid-based cytology was developed as an alternative to conventional cytology, which provides more homogeneous sampling of the cervical specimen (Hutchinson et al. 1994), and reduction in technical artifacts that limit conventional Pap test samples. However, recent systematic reviews conclude that because of the lack of well-designed comparative studies, convincing evidence to determine the superiority of the liquid based cytology for detecting high-grade lesions does not exist (Davey et al. 2006; Arbyn et al. 2008; Klinkhamer et al. 2003; Abulafia et al. 2003).

The variation in cell types each Pap smear image includes is another factor of complexity of these images. There are generally three types of squamous cells seen on Pap smear images (Fig. 2):

1. the superficial cells are the largest of the three types of cells and they have small pyknotic (degenerative) nuclei and cytoplasm that generally stains eosinophilic (red),
2. the intermediate squamous cells, which are similar in appearance but are slightly smaller in size and have larger, clearly structured, round nuclei with cytoplasm that usually stains basophilic (blue) and
3. the parabasal cell type that is smaller, more rounded and immature cell type.

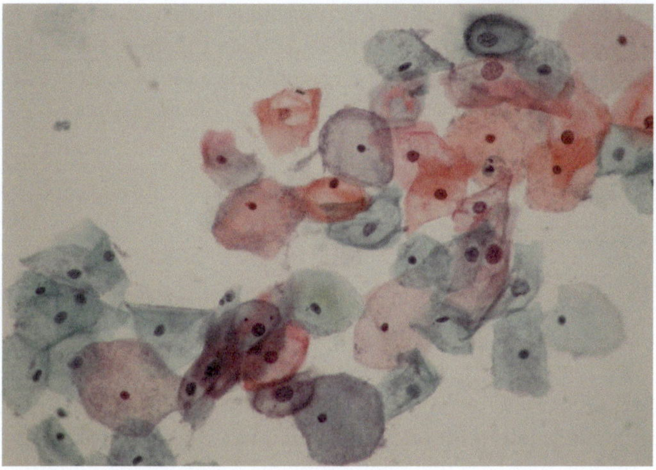

Fig. 2 The several types of cells that are included in Pap smear images (see text for details). In this image the high overlapping, the inhomogeneity in cells intensity and the existence of many artifacts are clearly present

In addition, the large number of cells (50,000–300,000 cells for an average slide), which must be reviewed by an expert, in order to characterize the smear as normal or abnormal is another important factor of difficulty. For all these reason, significant efforts and various research approaches and algorithms have been introduced focusing on the automated analysis of cytological Pap smear images.

In these images, the accurate detection of the nuclei is crucial because the nucleus is a very important structure within the cell and it presents significant changes when the cell is affected by a disease. In pathological situations, the nucleus may exhibit disproportionate enlargement, irregularity in form and outline, hyperchromasia or irregular chromatin condensation. The identification and quantification of these changes in the nucleus morphology and density contribute in the discrimination of normal and abnormal cells in Pap smear images. Thus, the challenge for any method proposed for the automated segmentation of cell nuclei is to overcome the limitations of the Pap smear images such as the high degree of cell overlapping, the lack of homogeneity in image intensity and the existence of many artifacts, in order to achieve accurate identification of the cells nuclei.

With the segmentation of the area of the cell, several salient features can be extracted, which can contribute to the classification of the cells and therefore the characterization of the smear slide. The existence of automated classification systems would not only diminish the required time for the smear classification, but also would avoid misclassifications due to human error. In the last years, several classification methods have been proposed for the Pap smear images, in order to identify abnormalities in the cells morphology. In general, these methods require that single cells in the slides are automatically isolated and analyzed.

In the next paragraphs, the segmentation and classification methods proposed for the cervical cell images are reported. The special features and the different scientific approach for each method are presented in detail in both cases. Furthermore, an attempt has been made for the methods that present similarities to be grouped under the same heading, in order to provide a more compact description for the general techniques that have been used in Pap smear images.

2 Segmentation of Pap Smear Images

The prerequisite for any further processing of Pap smear images and the derivation of conclusions for the characterization of their contents is the accurate determination of the cell nuclei area. However, the exact nuclei locations in the image are not clearly defined in many cases, mainly due to cell overlapping, inconsistent staining and the existence of many artifacts. Thus, the nuclei boundaries are quite ambiguous. For this reason, there are two open problems for every method proposed for the automated analysis of Pap smear images: the exact detection of nuclei locations and the accurate determination of nuclei boundaries. If a proposed method resolves the aforementioned restrictions, then it would be capable for direct application in images obtained from an optical microscope.

However, as these images are very complicated, many segmentation methods are applied in presegmented images containing a single cell or a part of the initial microscopic image (Fig. 3). The detection of the nuclei and cytoplasm boundaries in such images has been the subject of research for several scientists and many methodologies have been proposed. These methods exhibit remarkable performance in the segmentation of the structural parts of the cell. However the direct application of these methods in original Pap smear images, which may contain a large number of cells, cell overlapping and image artifacts would probably be doubtful, without an anterior step for the determination of the locations of the cells in the image.

Thus, more sophisticated approaches to the automated analysis of Pap smear images are the methods which are applied on images containing a large number of

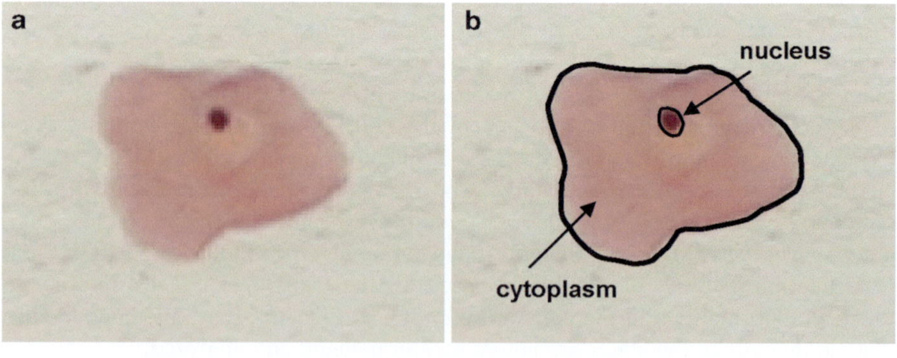

Fig. 3 (**a**) Original cell image. (**b**) The structural parts of the cell: the nucleus and the cytoplasm

isolated cells and cell clusters, which are clearly more complicated. These methods manage to exclude the background of the image and to recognize the locations and the boundaries of the cells. It must be noted that clustering algorithms are used in this scope for the rejection of false positive nuclei positions.

In the following paragraphs, the several detection and segmentation methods are categorized in terms of the general image processing method they are based on, for the determination of the nuclei boundaries in Pap smear images. In chronological order, the first approaches were quite simple and tested on limited data, while the most recent ones are more advanced and effective.

2.1 *Thresholding*

The first attempts to detect and segment cells in cervical microscopic images were based on image thresholding techniques exploiting the intensity histogram characteristics of the pixels (Cahn et al. 1977; Borst et al. 1979; Bengtsson et al. 1979). The earliest works were proposed in the late 1970s and this confirms the great scientific interest that these images attract. In general, the scope of these methods was to automatically identify threshold values in order to separate the cell from the background and the nucleus from the cytoplasm. A comparison of some threshold selection methods is included in MacAulay and Palcic (1988). In Poulsen and Pedron (1995), a method for the detection of regions of interest in reduced resolution Pap smear images is proposed. The segmentation problem is transformed into an optimisation process in Wu et al. (1998a), where the determination of an optimal threshold value refers to a parametric image, which is an approximation of the initial one. A thresholding technique was also proposed in Kim et al. (2007) for the binarization of the images and the determination of the nuclei locations, which entails in the feature extraction and the classification of the nuclei in normal or abnormal classes using a fuzzy RBF neural network. A multiscale local adaptive threshold method based on shape stability was proposed in Li and Najarian (2007) for the extraction of the nuclei region from the background, where the value of the threshold is obtained by preserving the expected shape of the segmented objects in the image. Finally, in Chang et al. (2009), the detection of the nuclei is accomplished through the thresholding of the images obtained after the application of mean filters and rescaling of the gray level of the pixels. A line-scan algorithm is also proposed for the detection of abnormal nuclei.

2.2 *Edge Detection*

Several methods developed for the segmentation of the nucleus and the cytoplasm are based on edge detection, which does not require any prior knowledge of the objects of interest in the image. Thus, in Tsai et al. (2008) a cytoplasm and nucleus contour (CNC) detector is developed, which segments the structural parts of cervical cells in presegmented images. The algorithm adopts the bi-group enhancer to suppress the noises and emphasize the object contours. Then, the K-means algorithm is used to discern the cytoplasm from the background and finally, the maximal colour difference (MCD) method can precisely determine the nucleus contour.

A similar semiautomatic method was proposed in Yang et al. (2008), where an edge enhancement nucleus and cytoplasm contour detector is presented and applied in presegmented cervical images. In a preprocessing step, three techniques, namely the trim-meaning filter, the bi-group enhancer and the mean vector difference enhancer are employed to eliminate the noise and enhance the contrast in cervical images. This results in the construction of an image having a bi-modal like intensity histogram, from which an optimal threshold can be obtained for the segmentation of the images. Finally, a relative-distance-error measure is presented to evaluate the segmentation error.

In Malm and Brun (2009) a nuclei segmentation algorithm that uses anisotropic dilation for curve closing is used. More specifically, the Canny edge detector followed by a series of morphological processes result in the detection of refined and closed edges. Then, in the edge image, a distance transform based on geodesic distances in the curved Riemmanian manifold is calculated, and a binary image is constructed after thresholding, skeletonization and morphological operations, resulting in the final segmentation of nuclei structures.

Another method based on edge detection for the segmentation of cervical images was proposed in Lin et al. (2009). At first, a colour depth equalization method is used in order to improve the contrast between the nucleus and the cytoplasm, and a Gaussian filter for noise removal. A two-group edge enhancement technique is then implemented based on the coarseness of each pixel. Then the Sobel operator and the non-maximum suppression are used for the extraction of the gradient image, which is binarized by setting an upper and lower edge value limit. Finally, the two longest closed curves from all the detected edges are selected to form the nucleus and the cytoplasm boundaries.

2.3 *Mathematical Morphology*

The powerful techniques that the mathematical morphology provides for image segmentation are exploited extensively in some published methods aiming at the analysis of Pap smear images. In Plissiti et al. (2011a) a method which mainly addresses the detection of nuclei locations in conventional Pap stained cervical cell images has been proposed. The method exploits the colour information as it searches for possible nuclei locations in all three channels of the colour image. For the final determination of the true nuclei set, local colour nuclei characteristics are used in a classification procedure. The method is not affected by the existence of overlapping cells and it can be applied to any cervical Pap smear image, which may contain both isolated cells and cell clusters. More specifically, the proposed method consists of four phases: (a) the preprocessing, (b) the detection of candidate cell nuclei centroids, (c) the refinement of candidate cell nuclei centroids and (d) the decision phase which includes the determination of the final nuclei locations. In the preprocessing phase, the regions of interest in the image occupied by the cells are defined. The outcome is a binary mask which indicates the cell clusters in the initial image. In the second phase, the identification of the probable locations of the centroids of the cells nuclei takes place. A morphological-based image process is proposed for the detection of the regional minima in the image, which indicate the existence of candidate nuclei of cells. The third phase is a procedure that exploits the spatial characteristics of the nuclei and the circumference of each nucleus is approximated, which results in the refinement of the nuclei centroids locations. In the fourth phase, a decision process which extracts the true nuclei locations is proposed and it is based on two steps. In the first step, an empirical rule which depends on the distance between the centroids is applied, for the reduction of false positive occurrences. In the second step, an unsupervised (fuzzy C-means) and a supervised (Support Vector Machine) classification techniques are used in order to determine the final set of nuclei centroids.

In Bamford and Lovell (1996) a water immersion algorithm, similar in spirit with watersheds was introduced for the detection of the location of isolated cells in a microscopic image. The method is composed of three stages: the quadtree smoothing, the lowest level classification and the boundary re-estimation. The outcome of this procedure is the determination of the boundary of the isolated cells. It must be noted that this method does not provide the detection of nuclei locations in cell clusters.

Methods based on watersheds for the analysis of Pap stained images have also been proposed in the literature. In Jackway (1996, 1995), images containing one single nucleus of a Pap stained squamous epithelial cell are oversegmented with the multiscale gradient watershed transform for the extraction of texture features, in order to characterize each cell as normal or abnormal. However, the problem of the detection of the accurate nuclei boundaries is not resolved.

In Plissiti et al. (2010a), and in more detail in Plissiti et al. (2011b), the detection of the candidate nuclei areas is based on a morphological image reconstruction process and the segmentation of the nuclei boundaries is accomplished with the application of the watershed transform in the morphological colour gradient image, using the nuclei markers extracted in the detection step. For the elimination of false positive findings (i.e. the detected regions that do not correspond to true nuclei areas), salient features characterizing the shape, the texture and the image intensity are extracted from the candidate nuclei regions and a classification step is performed to determine the true nuclei. The performance of two unsupervised (K-means, spectral clustering) and a supervised (Support Vector Machines, SVM) classification technique was examined, employing discriminative features which were selected with a feature selection scheme based on the minimum-Redundancy – Maximum-Relevance criterion. As it was concluded, the features with the most discriminative ability were those ones that concern not only the segmented region of the nuclei, but also the corresponding neighbourhood of the image.

Several other methods based on mathematical morphology have also been proposed. In Kim et al. (2006), sequential fuzzy greyscale morphological operations are performed for the extraction of the cell nucleus. The watershed transform has also been used in Kale and Aksoy (2010), where a hierarchical tree is implemented by performing multi-scale watershed segmentation on the image and then a region selection step using a measure based on spectral homogeneity and circularity identifies the regions occupied by cells in the image. In this method, a clustering step is performed, in which the cytoplasm and the nucleus regions are determined using an SVM. Finally, a method based on the watershed transform using multiscale morphological gradient and HSI colourspace was proposed in Nallaperumal and Krishnaveni (2008), and this method was also applied in macroscopic cancer images.

2.4 Pixel Classification

Another way of detecting the structural parts of the cells is based on pixel classification schemes. In Lassouaoui and Hamami (2003), a pixel classification method based on a multifractal algorithm is developed in a first step in order to classify the pixels of the image in background, cytoplasm or nucleus. Then, an optimization step is added, by learning through genetic algorithms, and the pixels are re-classified in the above three categories. This optimization step increases the precision of the borders of the detected areas, and in the same time decreases the confusion between various classes.

As the watershed transform has been extensively used for the segmentation of the nuclei boundaries in Pap smear images, salient markers of the nuclei locations were investigated. The method proposed by Lezoray and Cardot (2002) is based on pixel classification techniques for the detection of the nuclei markers, in order to avoid the oversegmentation that the watershed algorithm may produce. For this purpose, the K-means and a Bayesian classifier were used for the detection of the nuclei and other pixels class. For the later algorithm, a training set of images containing the ground truth was used for the estimation of the parameters of each Gaussian distribution.

Another approach that introduces a criterion function based on statistical structure of the objects in the image was proposed in Bak et al. (2004), which reflects both local and global characteristics of the image. A local spatial likelihood is defined and combined with local spatial prior probabilities, producing the local spatial posterior, which is treated as a criterion function. Initially, the pixels are clustered through the K-means algorithm, and then the segmentation of the image is achieved through an iterative procedure, in which each pixel is assigned at the most probable region (nucleus, cytoplasm or background) using the predefined criterion function.

In Sobrevilla et al. (2008), the areas of interest within the smears are detecting through fuzzy-based techniques and pixel classification. Then, the nuclei detection in images of higher magnification follows, which is also based on fuzzy rules.

Moreover, in Vaschetto et al. (2009), a three-color based algorithm that combines colour information, experts knowledge and fuzzy systems is proposed, aiming at the improvement of the accuracy of the method proposed in Sobrevilla et al. (2008) for the detection and the segmentation of the nuclei in Pap smear images.

Finally, a modified seed based region growing (MSBRG) algorithm for the automated segmentation of cervical cells was proposed in Mustafa et al. (2009). The K-means clustering algorithm classifies the pixels of the image into three categories (cytoplasm, nucleus and background) in a first step. Then from the extracted classification and using moments calculation, the location of seed pixels are determined and the MSBRG algorithm is performed. The method provides a simultaneous segmentation of all cells in the image, by identifying the pixels of the cytoplasm, the nucleus and the background.

2.5 Template Matching

Since the nucleus in general presents an ellipse-like shape, some methods based on template matching have been proposed, for the determination of the nuclei boundaries in Pap smear images. A parametric fitting algorithm for the segmentation of cell images with application to a cervical image was firstly proposed in Wu et al. (1998b), which employs both shape and regional image information. In this work, a parametric elliptical model for the nucleus shape is introduced, and the parameters are adjusted to fit the nucleus shape by minimizing a cost function. In order to find the optimal fitting, a coarse optimization search followed by a fine optimization procedure is performed. Thus, by incorporating the shape information, the number of parameters is significantly reduced and the method results in the detection of the nucleus boundary.

Furthermore, in Garrido and Perez de la Blanca (2000) a methodology based on a deformable template approach is developed. It consists of three steps: (1) initial estimation of the location of the cells in the image, (2) calculation of an elliptical approximation of the nucleus boundary and (3) refinement of the nucleus boundary using locally deforming models. More specifically, the detection of the location of the nuclei is obtained through a reformulation of the generalized Hough transform. Then the initial shapes of the nuclei are estimated by the determination of an ellipse. The final solution is obtained using Grenander's deformable template model, which converges in the correct nucleus boundary.

2.6 Deformable Models

In general, the application of deformable models for the definition of the nuclei boundary is constrained by the requirement for the initial approximation of the model, which is essential to be close to the real boundary of the nuclei in the image.

For this reason, the proposed methods based on deformable models have to solve also the problem of correct detection of the nucleus position in an image containing a large number of cells, or otherwise, their application is limited in presegmented images including only one cell.

Thus, in Bamford and Lovell (1998), the boundary of the nuclei is determined on presegmented images with the use of an active contour model, which is initialized through the construction of a search space, consisting of two concentric cycles. In the limits of the predefined search space, the most probable locations for the points of the nuclei boundaries are defined, following a Viterbi search-based dual active contour algorithm.

Moreover, a segmentation method combining the physically based models and the active shape models was presented in Plissiti and Nikou (2011). Thus, a compact representation of the shape of the nucleus is provided and a-priori knowledge of the expected nucleus shape is exploited. An efficient framework for the training of active shape models (ASM), representing smooth shapes, which is based on the representation of a shape by the vibrations of a spring-mass system is described. A deformable model whose behavior is driven by physical principles is used on a training set of nuclei. The boundaries of the nuclei of the training set are detected with the convergence of the physics-based deformable model and attributes of the shapes of the nuclei are expressed in terms of modal analysis. Based on the estimated modal distribution, the modal amplitudes of the learnt model are used and the cost function to be minimized is accordingly modified for the detection and description of an unknown nucleus in a new image. The proposed method is evaluated using cytological images of conventional Pap smears. The method has been tested in terms of the accurate segmentation of the nuclei borders in images from Pap smear slides, and as it was verified by the results it presents high performance.

The above methods were applied in presegmented nuclei images. A more general method which does not require presegmented images was presented in Plissiti et al. (2010b). The method consist of three phases: (a) the definition of a candidate nuclei centroids set using mathematical morphology, (b) the initial approximation of cells nuclei boundaries and (c) the application of Gradient Vector Flow (GVF) snakes for the final estimation of candidate cell nuclei boundaries. It must be noted that the initial approximation of each snake position is obtained automatically, without any user interference. For the final determination of the nuclei in the images, fuzzy C-means clustering, using a data set of patterns based on the characteristics of the area enclosed by the final position of the Gradient Vector Flow snakes was performed.

In Harandi et al. (2010), a method for the localization of cells in low resolution combined with the detection of the nucleus and cytoplasm boundaries in high resolution is proposed. The geometric active contour without re-initialization (GACWR) method is used for the localization of the cervical cells in an image of low resolution, which are then classified in free-lying cells, connected cells and irrelevant objects. After the detection of the position of each cell, the original image is partitioned into subimages containing the detected cells, and on each subimage, a binary mask is generated, in which the unwanted objects elimination and the selection of the objects of interest is achieved. In order to separate the different cells in the cell clusters, the cells are firstly modelled by a circle, which serves as the initial contour of a GACWR model. It must be noted that the GACWR is applied on an image of higher resolution, and entails in the identification of the cytoplasm. The same procedure is followed for the delineation of the nuclei boundaries, which is also initially approximated by a circle.

2.7 Segmentation of Overlapping Nuclei in Cervical Images

The first work concerning the recognition of overlapping cells nuclei in Pap smear images was proposed in Bengtsson et al. (1981). The overlap detection algorithm was based on information both from the nucleus contour and from the density profile. The method was based on the detection of significant concavities along the contour and the extraction of a set of features for the classification of the object of interest as single or overlapping. It must be noted that the nucleus contour is extracted through a thresholding procedure.

In more recent studies, the distance transform in a binary image containing the regions of the nuclei is calculated in Jung et al. (2010) and the topographic surface generated by the distance transform is considered as a Gaussian mixture. The EM algorithm is then applied for the determination of the parameters of each nucleus cluster and the separation line of the overlapped nuclei is defined with the minimization of a criterion function. It must be noted, that in this approach, the occluded area of each nucleus is reconstructed with a constrained ellipse fitting technique.

In Jung and Kim (2010), the segmentation of clustered nuclei is treated as an optimization problem and a marker extraction scheme based on the H-minima transform is introduced to obtain the optimal segmentation result from the distance map.

3 Classification of Pap Smear Images

The examination of the Pap smear images by expert cytologists results in the characterization of the slide, and in many cases the specific disease of the cells can be recognized. The classes that a smear slide can automatically be classified by an expert system are determined by a specialist physician. For the creation of a standardized framework for laboratory reports, the Bethesda System (Solomon et al. 2001) was developed. This system was firstly introduced in 1989 and updated in 2001 as a uniform system of terminology that would provide clear guidance for clinical management. In this system, the smears are divided in two categories: normal and abnormal. The abnormal category can be further divided into four groups: atypical squamous cells (ASC), low grade squamous intraepithelial lesion (LSIL), high grade squamous intraepithelial lesion (HSIL) and squamous cell carcinoma.

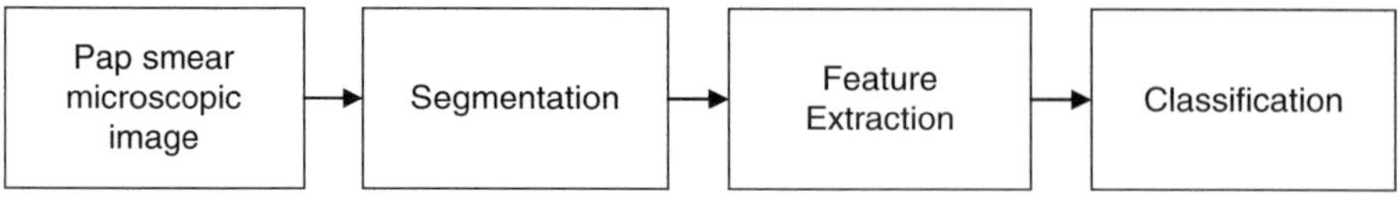

Fig. 4 Schema of Pap smear microscopic image classification process

Furthermore, in Jantzen et al. (2005) a Pap-smear benchmark database is presented. The database consists of 917 samples distributed unevenly in seven classes by the careful examination of skilled cyto-technicians and doctors. There are two main classes that a cell may be classified: normal and abnormal. Both of these classes can be further separated into more categories. Thus, superficial squamous epithelial, intermediate squamous epithelial and columnar epithelial cells constitute the normal class. On the other hand, the characterization of the cells as mild squamous non-keratinizing dysplasia, moderate squamous non-keratinizing dysplasia, severe squamous non-keratinizing dysplasia, and squamous cell carcinoma in situ intermediate indicate the existence of abnormality in the Pap smear slide.

The characterization of the slide is mainly feasible by the evaluation of salient features of the nuclei of the cells, which, in general, is subjective to the experience of the observer. In the last years, some efforts have been made in order to obtain reproducible and objective characterization of the Pap smear images through computer vision, reducing the dependency to human experts. In this scope, several commercial interactive systems have been developed, aiming at the automated classification of the Pap smear images. These systems, such as PAPNET, AutoPap and TriPath are based on machine learning algorithms in order to assist in cervical cytology screening.

Besides the commercial packages, the automated classification of Pap smear images has been an interesting field for the researchers and many methods have been proposed, which involve both intelligent feature extraction techniques and machine learning algorithms, in order to recognize abnormalities in these images. The basic steps of these methods are depicted in Fig. 4. In the following paragraphs, the features extracted from Pap smear images and the proposed techniques for their classification are described.

3.1 Features of Cervical Cells Used for Classification

The published methods which deal with the classification of Pap smear images are based on the calculation of features extracted from the areas of the nucleus and the cytoplasm. It must be noted that most of these methods use presegmented images which contain only one cell, so the correct segmentation of the nucleus and the cytoplasm is feasible. In images containing cell clusters, the detection of the cytoplasm boundary is a difficult problem and until now, there is not any method in the literature that results in the automated delineation of cytoplasm in cell clusters.

The features used in these methods involve both intensity and shape characteristics of the nucleus and the cytoplasm. The most commonly used shape features are the area (size), the nucleus to cytoplasm area ratio, the shortest and longest diameter, the perimeter, the elongation and the roundness (circularity). Furthermore, the positions of the nucleus and cytoplasm centroids are used for the definition of the relative placement of the nucleus inside the cytoplasm.

As far as the intensity characteristics are concerned, the brightness, the mean grey level and intensity disparity between the nucleus and the background are used by several techniques. Moreover, some textural features such as the uniformity, entropy, smoothness, moments and local binary patterns have also been proposed Plissiti et al. (2011b).

In many methods, more sophisticated features are calculated in Pap smear images. Thus, in Chankong et al. (2009) a set of features extracted from the two dimensional Fourier transform of single cell images is utilized, without the need of a segmentation step, which would result in the detection of the cytoplasm and nuclei boundaries. The proposed features are based on the mean, the variance and the entropy obtained from the frequency components along the circle having as center the center of the spectrum and the frequency components along the radial line having an angle θ.

Feature vectors based on the wavelet transform were proposed in Bondarenko and Katsuk (2005). The statistical wavelet analysis and the wavelet analysis in "Brute force" approach are combined for the extraction of feature patterns. These features are then used as input in a neural network classifier for the final classification of the image in normal or abnormal categories.

Several methods of automated feature selection have been proposed for the construction of subsets of features with high discriminative ability. Thus in Marinakis et al. (2009), a technique based on a genetic algorithm is proposed for the selection of the best optimal performing feature subset. The algorithmic scheme is combined with a number of nearest neighbour based classifiers, and a comparison with a tabu search based metaheuristic algorithm for feature selection is presented. Furthermore, in Marinakis et al. (2008), a method based on particle swarm optimization is proposed for the feature selection problem. In this work, nearest neighbour techniques were also used for the calculation of the performance of the proposed algorithm. The results were obtained for both the two-class problem (normal or abnormal) and for the seven-class problem (as they were described in the previous section).

3.2 Classification Techniques

The extracted features are used for training classifiers, in order to recognize and classify a certain image/cell into the corresponding class. The first attempts to classify the cells in Pap smear images were based on the Bayes rule. Thus, in Oliver et al. (1979) a training set containing normal and abnormal cells is used for the calculation of the parameters of the classifier, and two decision rules are combined for the classification of each cell. In Walker et al. (1994), the parametric Bhattacharyya distance is used for the determination of a pair of textural features and the Bayes classifier is then applied for the classification of the samples. The Bayesian classifier was also used in Chankong et al. (2009) and Riana and Murni (2009).

Another widespread technique used for the classification of Pap smear images are the artificial neural networks. In Mat Isa et al. (2008), a hierarchical hybrid multilayer perceptron network (H^2MLP) is illustrated for the classification of cervical images into three categories: normal, low-grade squamous intraepithelial lesion (LSIL) and high grade squamous intraepithelial lesion (HSIL). In this work, the input data are first classified as normal and abnormal by a hybrid multilayer perceptron (HMLP), and in the abnormal case, they are further classified by a second HMLP into the LSIL and HSIL categories. The performance of the H^2MLP neural network has been compared with RBF, MLP and HMLP artificial neural networks with the number of hidden nodes varying from 1 to 50. The effectiveness of the proposed diagnostic system has been demonstrated using 550 annotated Pap smear images.

The Rank M-Type Radial Basis Function (RMRBF) neural network was implemented in Gallegos-Funes et al. (2009) for the classification of microscopic Pap smear images. The Median M-Type (MM) estimator was used in an RBF neural network for the estimation of the parameters of the proposed network. The results presented in terms of the sensitivity and specificity were compared with the results of simple RBF and median RBF neural network.

Several other proposed methods are based on neural networks. A feed forward neural network with a single hidden layer is used in Bondarenko and Katsuk (2005) and the training method selected was backpropagation with a variable learning rate. A multilayer sigmoid neural network along with Levenberg-Marquardt backpropagation training algorithm was implemented in Li and Najarian (2001) for the classification of samples for which fuzzy based classification is unclear.

Furthermore, some classification techniques based on fuzzy logic were also proposed. Thus, in Kim et al. (2007), the fuzzy RBF network is applied to classify and identify the normal and abnormal nuclei. In Kim et al. (2006), the fuzzy Adaptive Resonance Theory (ART) algorithm is used in order to classify and distinguish normal, abnormal and cancerous cells, based on the standard categories of the Bethesda System.

Support Vector Machines (SVM) were also used for the classification of Pap smear cells into normal and abnormal categories (Chankong et al. 2009; Huang et al. 2007). It must be noted that SVM have been also used for the automated detection of cell nuclei (Plissiti et al. 2010a, 2011a) in cervical images captured directly from an optical microscope, which contain overlapping cells and cell clusters. Moreover, a method for cervical cancer detection using SVM based feature screening has also been proposed (Zhang and Liu 2004), and it concerns the reduction of the feature space dimension in multispectral Pap smear images.

4 Conclusion

The task of the automated analysis and classification of Pap smear images is one of the most interesting and challenging issues in computer vision and artificial intelligence scientific fields. The identification of cervical cell nuclei areas in conventional Pap smear images is a difficult problem, as these images present great complexity and certain limitations. The accurate segmentation of the area of nucleus is a prerequisite for the derivation of diagnostic conclusions and the characterization of the contents in Pap smear images. This is feasible with the extraction of salient features which contribute in the discrimination of the cells in normal and abnormal categories by the expert classifiers. As we can conclude, great effort has been made by several researchers in order to present effective techniques concerning both the segmentation and the classification of Pap smear images. These techniques have been reviewed by the current study, in order to provide an integrated report of the state of the art methods for the automated analysis and classification of Pap smear images. Although the proposed techniques present high performances, the automated processing of a Pap smear slide, which would result in the documented diagnosis is not possible yet. In the future, the development of fully automated methods for the Pap smear interpretation is expected.

References

Abulafia O, Pezzullo JC, Sherer DM (2003) Performance of ThinPrep liquid-based cervical cytology in comparison with conventionally prepared Papanicolaou smears. Gynecol Oncol 90(1):137–144

Arbyn M, Bergeron C, Klinkhamer P, Martin-Hirsch P, Siebers AG, Bulten J (2008) Liquid compared with conventional cervical cytology: a systematic review and meta-analysis. Obstet Gynecol 111(1):167–177

Bak E, Najarian K, Brockway JP (2004) Efficient segmentation framework of cell images in noise environments. Proc 26th Int Conf IEEE Eng Med Biol 1:1802–1805

Bamford P, Lovell B (1996) A water immersion algorithm for cytological image segmentation. In: Proceedings of the APRS image segmentation workshop, Sydney, pp 75–79

Bamford P, Lovell B (1998) Unsupervised cell nucleus segmentation with active contours. Signal Process 71(2):203–213

Bengtsson E, Eriksson O, Holmquist J (1979) High resolution segmentation of cervical cells. J Histochem Cytochem 27(1):621–628

Bengtsson E, Eriksson O, Holmquist J, Jarkrans T, Nordin B, Stenkvist B (1981) Segmentation of cervical cells: detection of overlapping cell nuclei. Comput Graph Image Process 16:382–394

Bondarenko AN, Katsuk AV (2005) Extracting feature vectors of biomedical images. In: Proceeding of the 9th Russian-Korean international symposium science and technology KORUS 2005 (Bondarenko and Katsuk) is Novosibirsk, 26 June–2 July 2005, Russia

Borst H, Abmayr W, Gais P (1979) A thresholding method for automatic cell image segmentation. J Histochem Cytochem 27(1):180–187

Cahn RL, Poulsen RS, Toussaint G (1977) Segmentation of cervical cell images. J Histochem Cytochem 25(7):681–688

Chang CW, Lin MY, Harn HJ, Harn YC, Chen CH, Tsai KH, Hwang CH (2009) Automatic segmentation of abnormal cell nuclei from microscopic image analysis for cervical cancer screening. In: Proceeding of the 3rd IEEE international conference on nano/molecular medicine and engineering, Tainan, Taiwan, pp 77–80

Chankong T, Theera-Umpon N, Auephanwiriyakul S (2009) Cervical cell classification using Fourier transform. Proc 13th Int Conf Biomed Eng 23:476–480

Davey E, Barratt A, Irwig L, Chan SF, Macaskill P, Mannes P, Saville AM (2006) Effect of study design and quality on unsatisfactory rates, cytology classifications, and accuracy in liquid based versus conventional cervical cytology: a systematic review. Lancet 367(9505):122–132

Gallegos-Funes FJ, Gomez-Mayorga ME, Lopez-Bonilla JL, Cruz-Santiago R (2009) Rank M-type radial basis function (RMRBF) neural network for Pap smear microscopic image classification. Apeiron 16(4):542–554

Garrido A, Perez de la Blanca N (2000) Applying deformable templates for cell image segmentation. Pattern Recognit 33(5):821–832

Harandi NM, Sadri S, Moghaddam NA, Amirfattahi R (2010) An automated method for segmentation of epithelial cervical cells in images of ThinPrep. J Med Syst 34:1043–1058

Huang PC, Chan YK, Chan PC, Chen YF, Chen RC, Huang YR (2007) Quantitative assessment of Pap smear cells by PC-based cytopathologic image analysis system and support vector machine. Proc 1st Int Conf Med Biometrics Lect Notes Comput Sci 4901:192–199

Hutchinson ML, Isenstein LM, Goodman A, Hurley AA, Douglass KL, Mui KK, Patten FW, Zahniser DJ (1994) Homogeneous sampling accounts for the increased diagnostic accuracy using the ThinPrep processor. Am J Clin Pathol 101(2):215–219

Jackway P (1995) Morphological multiscale gradient watershed image analysis. In: Proceedings of the 9th Scandinavian Conference Image Analysis (SCIA 1995), Uppsala, pp 87–94

Jackway P (1996) Gradient watersheds in morphological scale space. IEEE Trans Image Process 5:913–921

Jantzen J, Norup J, Dounias G, Bjerregaard B (2005) Pap-smear benchmark data for pattern classification. In: Proceedings of the Nature inspired Smart Information Systems (NiSIS), Albufeira, pp 1–9

Jung C, Kim C (2010) Segmenting clustered nuclei using H-minima transform-based marker extraction and contour parameterization. IEEE Trans Biomed Eng 57(10):2600–2604

Jung C, Kim C, Wan Chae S, Oh S (2010) Unsupervised segmentation of overlapped nuclei using Bayesian classification. IEEE Trans Biomed Eng 55(12):2825–2832

Kale A, Aksoy S (2010) Segmentation of cervical images. In: Proceedings of the 20th international conference on pattern recognition, Istanbul, pp 2399–2402

Kim KB, Kim S, Sim KB (2006) Nucleus classification and recognition of uterine cervical Pap-Smears using fuzzy ART algorithm. Proc 6th Int Conf Simul Evol Learning, Lect Notes Comput Sci 4247:560–567

Kim KB, Song DH, Woo YW (2007) Nucleus segmentation and recognition of uterine cervical Pap-smears. Proc 11th RSFDGrC(2007), Lect Notes Comput Sci 4482:153–160

Klinkhamer PJ, Meerding WJ, Rosier PF, Hanselaar AG (2003) Liquid based cervical cytology: a review of the literature with methods of evidence-based medicine. Cancer Cytopathol 99 (5):263–271

Lassouaoui N, Hamami L (2003) Genetic algorithms and multifractal segmentation of cervical cell images. Proc 7th Int Symp Signal Processing Appl 2:1–4

Lezoray O, Cardot H (2002) Cooperation of colour pixel classification schemes and colour watershed: a study for microscopic images. IEEE Trans Image Process 11(7):783–789

Li Z, Najarian K (2001) Automated classification of Pap smear tests using neural networks. Proc Int Joint Conf Neural Networks. Washington, Dc 4:2899–2901. Accessed 15–19 July 2001

Li Z, Najarian K (2007) Biomedical image segmentation based on shape stability. In: Proceedings of the 14th IEEE International Conference on Image Process (ICIP 2007), San Antonio, pp 281–284

Lin CH, Chan YK, Chen CC (2009) Detection and segmentation of cervical cell cytoplast and nucleus. Int J Imag Syst Technol 19(3):260–270

MacAulay C, Palcic B (1988) A comparison of some quick and simple threshold selection methods for stained cells. Anal Quant Cytol Histol 10(2):134–138

Malm P, Brun A (2009) Closing curves with Riemannian dilation: application to segmentation in automated cervical cancer screening. Proc 5th Int Symp Visual Comput 2009, Lect Notes Comput Sci 5875:337–346

Marinakis Y, Marinaki M, Dounias G (2008) Particle swarm optimization for Pap-smear diagnosis. Expert Syst Appl 35:1645–1656

Marinakis Y, Dounias G, Jantzen J (2009) Pap smear diagnosis using a hybrid intelligent scheme focusing on genetic algorithm based feature selection and nearest neighbour classification. Comput Biol Med 39:69–78

Mat Isa NA, Mashor MY, Othman NH (2008) An automated cervical pre-cancerous diagnostic system. Artif Intell Med 42:1–11

Mustafa N, Mat Isa NA, Mashor MY (2009) Automated multicells segmentation of ThinPrep® image using modified seed based region growing algorithm. Biomed Soft Comput Hum Sci 14 (2):41–47

Nallaperumal K, Krishnaveni K (2008) Watershed segmentation of cervical images using multiscale morphological gradient and HSI colour space. Int J Imag Sci Eng (IJISE) 2 (2):212–216

Oliver LH, Poulsen RS, Toussaint GT, Louis C (1979) Classification of atypical cells in the automatic cytoscreening for cervical cancer. Pattern Recognit 11:205–212

Papanicolaou GN (1942) A new procedure for staining vaginal smears. Science 95(2469):438–439

Plissiti ME, Nikou C (2011) Cell nuclei segmentation by learning a physically based deformable model. In: Proceedings of the 17th international conference on Digital Signal Processing (DSP'11) Corfu, Greece. Accessed 6-8 July 2011

Plissiti ME, Nikou C, Charchanti A (2010a) Watershed-based segmentation of cell nuclei boundaries in Pap smear images. In: Proceedings of the 10th IEEE international conference on Information Technology Applications in Biomedicine (ITAB'10), Corfu, Greece. Accessed 3–5 November 2010

Plissiti ME, Nikou C, Charchanti A (2010b) Accurate localization of cell nuclei in pap smear images using gradient vector flow deformable models. In: Proceeding of the 3rd international conference on Bio-inspired Signals and Systems (BIOSIGNALS), Valencia, Spain, pp 284–289. Accessed 20–23 January 2010

Plissiti ME, Nikou C, Charchanti A (2011a) Automated detection of cell nuclei in pap smear images using morphological reconstruction and clustering. IEEE Trans Inf Technol Biomed 15(2):233–241

Plissiti ME, Nikou C, Charchanti A (2011b) Combining shape, texture and intensity features for cell nuclei extraction in pap smear images. Pattern Recognit Lett 32:838–853

Poulsen RS, Pedron I (1995) Region of interest finding in reduced resolution colour imagery-application to cancer cell detection. Pattern Recognit 28(11):1645–1655

Riana D, Murni A (2009) Performance evaluation of Pap smear cell image classification using quantitative and qualitative features based on multiple classifiers. In: Proceedings of the international conference on Advanced Computer Science and Information Systems (ACSIS'09) Jakarta, Indonesia pp 38–42. Accessed 7–8 December 2010

Sobrevilla P, Montseny E, Lerma E (2008) A fuzzy-based automated cell detection system for color Pap smear tests –FACSDS. In: Fuzzy sets and their extensions: representation, aggregation and models, studies in fuzziness and soft computing. Springer Berlin, Heidelberg, 220:657–674

Solomon D, Davey D, Kurman R, Moriarty A, O'Connor D, Prey M, Raab S, Sherman M, Wilbur D, Wright T, Young N (2001) The 2001 Bethesda system: terminology for reporting results of cervical cytology. J Am Med Assoc 287:2114–2119

Tsai MH, Chan YK, Lin ZZ, Yang-Mao SF, Huang PC (2008) Nucleus and cytoplast contour detector of cervical smear image. Pattern Recognit Lett 29:1441–1453

Vaschetto F, Montseny E, Sobrevilla P, Lerma E (2009) THREECOND: An automated and unsupervised three colour fuzzy-based algorithm for detecting nuclei in cervical Pap smear

images. In: Proceedings of the 9th international conference on intelligent systems design and applications, Washington DC, pp 1359–1364
Walker RF, Jackway P, Lovell B, Longstaff ID (1994) Classification of cervical cell nuclei using morphological segmentation and textural feature extraction. In: Proceedings of the 2nd Australian and New Zealand conference on intelligent information systems, Brisbane, pp 297–301
Wu HS, Gil J, Barba J (1998a) Optimal segmentation of cell images. Proc IEE Vision,Image Signal Process 145(1):50–56
Wu HS, Barba J, Gil J (1998b) A parametric fitting algorithm for segmentation of cell images. IEEE Trans Biomed Eng 45(3):400–407
Yang-Mao SF, Chan YK, Chu YP (2008) Edge enhancement nucleus and cytoplast contour detector of cervical smear images. IEEE Trans Syst Man Cybernet B-Cybernet 38(2):353–366
Zhang J, Liu Y (2004) Cervical cancer detection using SVM based feature screening. In: Proceeding of the 7th international conference on Medical Image Computing and Computer Aided Intervention (MICCAI'04), Saint-Malo, France pp 873–880

Modeling of Trabecular Architecture as Result of an Optimal Control Procedure

Ugo Andreaus, Michele Colloca, and Daniela Iacoviello*

Abstract The primary mechanical function of bones is to provide rigid levers for muscles to pull against and to remain as light as possible to allow for efficient locomotion. To accomplish this function, bones must adapt their shape and architecture efficiently. Bone tissue during skeletal growth and development continuously adjusts its mass and architecture to changing mechanical environments.

The trabecular structure in bone is the result of a dynamic remodeling process controlled by mechanical loads.

In this study, the process of adaptive bone remodeling is investigated mathematically and simulated by a finite element model. Bone tissue is described as continuous material with variable mass density. Topology optimization and cellular automata models are adopted with the aim to characterize the osteocytes located within the bone as sensors of mechanical signals.

An implemented optimization process provides structures resembling actual trabecular architectures and aligning them with the actual principal stress orientation. Two cost indices are derived from the structural optimization task of simultaneously minimizing the weight and maximizing the stiffness of the investigated domain.

---

*All the authors equally contributed to this work

U. Andreaus (✉)
Department of Structural and Geotechnical Engineering, Faculty of Civil and Industrial Engineering, Sapienza University of Rome, Via Eudossiana 18, 00184 Rome, Italy
e-mail: ugo.andreaus@uniroma1.it

M. Colloca
Department of Mechanical and Aerospace Engineering, Polytechnic Institute of New York University, Six MetroTech Center, Brooklyn, NY 11201, USA
e-mail: mcolloca@poly.edu

D. Iacoviello*
Department of Computer, Control and Management Engineering Antonio Ruberti, Sapienza University of Rome, Via Ariosto 25, 00185 Rome, Italy
e-mail: daniela.iacoviello@dis.uniroma1.it

D. Iacoviello and U. Andreaus (eds.), *Biomedical Imaging and Computational Modeling in Biomechanics*, Lecture Notes in Computational Vision and Biomechanics 4,
DOI 10.1007/978-94-007-4270-3_2,

A trabecular architecture characterized by the highest structural efficiency is showed as result of the proposed optimal control process.

Keywords Optimal control • PID control • Finite element method (FEM) • Bone functional adaptation • Minimum mass • Maximum stiffness

1 Introduction

Living bone is continually undergoing processes of growth, reinforcement, and resorption that are collectively termed *bone remodeling* or *functional adaptation*. The functional adaptation process in living bone means that the bone adapts its overall structure to changes in its load environment. Understanding and predicting the adaptation properties of living bone are particularly important for proper design of prosthetic devices in contact with bone e.g. fracture fixation plates, surgical screws, and artificial joints (Huiskes et al. 1987). If orthopaedic implants place a different stress from the one that the bone tissue is accustomed to, the functional adaptation is activated. The remodeled bone tissue could be weaker and lead to failure of the surgical procedure. It has been suggested that femoral resorption could be the predominant factor in implant stem fracture of a total hip replacement (Cowin and Doty 2007).

The objective of a theory for tissue adaptation is to provide numerical models or boundary value problems that predict the adaptive behaviour of the tissue under altered environmental load (Cowin and Doty 2007). Bone remodeling formulations, for the most part, fail into two categories: the first one collects models of local regulation processes at the tissue and cellular level (Cowin and Hegedus 1976; Fyhrie and Carter 1986; Carter 1987; Huiskes et al. 1987; Beaupré et al. 1990a; Turner 1991); the second one refers to models which assume global optimality of the bone structure.

The first published work arguing that bone is globally optimally suited to support the loads to which it is exposed with a minimum amount of material has been credited to Bourgery (1832) by Roesler (1987). Later, this idea was adopted by Roux (1881) and Wolff (1892) as well. Recent investigators have also proposed a maximum-minimum approach (Curry 1984; Fyhrie and Carter 1986; Subbarayan and Bartel 1989). Jacobs et al. (1997) introduced a mathematical approach combining any existing rule for mechanobiologic adaptation of mass density and assuming that the cancellous bone is mechanically optimized to support the mechanical load. Bagge (2000) derived the remodeling rate equation from the structural optimization task of maximizing the stiffness in each time step; the remodeling equilibrium parameter, meaning that no deposition or resorption takes place, was defined as the element optimality conditions and the optimal design was used as the initial design for the onset of the remodeling simulation.

Lekszycki (2005) made use of the hypothesis of optimal response of bone and derived (instead of postulating) the remodeling rules from a very general and global assumption.

Tovar et al. (2006) used the hybrid cellular automaton (HCA) algorithm to incorporate a distributed control loop within a structure where ideally localized sensor cells activate local processes of material growth and resorption. With a specific control strategy, the overall structure was driven to an optimal configuration.

In many theoretical models the concept of an error signal driving the adaptation process is used. Implicit in the use of an error-driven theory is the assumption that there exists some remodeling equilibrium condition in which the mechanical environment of bone tissue is suitably adapted without undergoing net deposition or resorption. Only deviations from this remodeling equilibrium condition would initiate net remodeling activity in the mechanical environment. Some authors have used a single value to represent the mechanical remodeling equilibrium state; others have used the concept of a minimum signal or a lazy zone to represent a range of values for which the bone would also be quiescent (Frost 1983; Carter 1984; Huiskes et al. 1987).

Moreover, relationships between local bone apparent density and loading history are developed by assuming that bone mass is adjusted in response to energy considerations (Carter et al. 1987, 1989). In adaptive remodeling models, strain energy density (*SED*) is used as a feed-back control variable to determine shape or bone density adaptation to alternative functional requirements, where by homeostatic *SED* distribution is assumed as the remodeling objective (Huiskes et al. 1987).

In Tovar et al. (2006) the assumed remodeling rule was based on the classical proportional–integral–derivative (PID) control, applied to the error signal between the strain energy density and a target. In Liu and Daley (1999) the problem of PID controller tuning was addressed in an optimization framework in the frequency domain; the considered performance functions were the gain margins, the phase margin and the steady-state error. In Liu and Daley (2001) different design schemes were proposed adopting time-domain and frequency-domain optimal tuning PID control and multiobjective optimal-tuning PID control. In Lennartson and Kristansson (2006) an analytical method was proposed and related to multi-criteria optimization problem and optimal PID controller evaluated with and without *Smith predictor*.

In this paper, the implementation of a new method for optimal control is proposed to model trabecular architecture of a Michell-type two-dimensional structure as a continuum design domain. The new procedure was originally developed by the authors to predict optimal configurations of internal architecture in cortical bone samples (Andreaus et al. 2011, 2012). The optimization problem consists of maximizing the stiffness of the structure and minimizing its mass. Two alternative evolutions of material parameters, namely mass density and elastic modulus, are considered, and two cost indices, defined as linear combination and product of total mass and strain energy, are minimized and investigated. Furthermore, the classical PID control rule is applied to the error signal between the strain energy density and a physiological target.

2 Evolution Rules

In bone tissue osteocytes can detect a mechanical stimulus that causes local material adaptations. This detection-activation process can be described with a generic mathematical expression, using the apparent density as characterization of the internal morphology. The rate of bone apparent density change $d\rho/dt$, with $\rho = \rho(x,y,z)$ at a particular location $P(x,y,z)$, can be described as an evolution function F, which depends on a particular stimulus at that location $P(x,y,z)$. It is assumed that the stimulus is directly related to the local mechanical load in the bone and can be determined from the local stress tensor $\sigma_{ij}(x,y,z)$, the local strain tensor $\varepsilon_{ij}(x,y,z)$, and the apparent density $\rho = \rho(x,y,z)$ by:

$$\frac{d\rho}{dt} = F\left(\sigma_{ij}, \varepsilon_{ij}, \rho\right) \quad 0<\rho<\rho_{max} \tag{1}$$

where ρ_{max} is the maximum density of cortical bone. When the evolution function F reaches zero, the system is in equilibrium and the rate of bone density change is zero.

The generic relationship (1) can be specified as:

$$\frac{d\rho}{dt} = c_\rho(S - S^*) \quad 0<\rho<\rho_{max} \tag{2}$$

where c_ρ is a constant, $S = S(x,y,z)$ is the mechanical stimulus and S^* (x,y,z) is a reference value. If the bone sample is discretized into “n” cells (cellular automata CAs), the stimulus S is usually expressed per cell. In other words, there is precisely one sensor point per one cellular automaton.

Equation 2 means that the stimulus strives to become equal to the reference value S^*, which can either be site-specific (the reference state of the stimulus depends on location) or non-site-specific (the reference state of the stimulus is constant). This implicates that the normal stimulus distribution $S = S^*(x,y,z)$ must be known or determined from a normal equilibrium density distribution, in order to predict the bone adaptation process to an abnormal situation, e.g. after loading changes, geometry or implant placement. Estimates could be derived from investigations by Rubin (1984) and Rubin and Lanyon (1987) who found that peak values of strains in bone of several species during physiological activities vary between 2,000 and 3,000 μstrain. If uniaxial strain state is assumed, with a tissue modulus of 5,000 MPa, normal values of peak strain energy density vary from 0.01 to 0.0225 MPa. This is the range of variation of the reference value, SED^*, used by Mullender and Huiskes (1995). In this study, the strain energy density (*SED*) defined as:

$$SED = \frac{1}{2}\sigma_{ij}\varepsilon_{ij} \tag{3}$$

is used as a feed-back control variable for adaptive remodeling. This choice is well-supported on physical grounds (Carter et al. 1987). The strain energy is a physical quantity related to both rigidity and stress. From a methodological standpoint it has the advantage of being a scalar variable and an invariant of the strain tensor, which implies that only one remodeling coefficient (c_ρ) must be established (see Eq. 2). Thus, the difference between the local actual $SED(x,y,z;t)$ and a non-site-specific homeostatic equilibrium SED^*, is assumed as the driving force for adaptive activity. When $SED > SED^*$ or $SED < SED^*$, adaptive activity initiates and the rate of adaptation is proportional to c_ρ.

Consequently, the changes in the apparent density ρ in each CA are converted to changes in elastic modulus by using:

$$E(x,y,z;t) = C\,[\rho(x,y,z;t)]^{\gamma} \tag{4}$$

where C and γ are constants. As the material density changes during remodeling process, the elastic modulus is calculated as a power function of the bone tissue apparent density ρ in each location $P(x,y,z)$. The relationship between density and Young's modulus is taken from Carter and Hayes (1977). The use of a cubic power relationship is in accordance with experimental data from Currey (1988).

The basic idea of the proposed algorithmic method was to begin with a model in which each cell had a uniform density distribution and then to apply loads and iterate the updating of the density value in each CA through a remodeling rule. In the procedure which follows, the bone tissue was initially considered as isotropic and homogeneous. The apparent density and elastic modulus subsequently varied as a function of position as the functional adaptation process remodeled the bone. The rate of adaptation was proportional to the difference $SED\text{-}SED^*$, according to the proportionality constant c_ρ:

$$\frac{d\rho_{\mathrm{i}}}{dt} = c_\rho (SED_{\mathrm{i}} - SED^*) \tag{5}$$

for i = 1,2....n, where "n" is the number of cellular automata in which the bone is discretized and SED_i is the strain energy density of the CA at the discrete location i. For this method there was no one-to-one correspondence between CA lattice and the FE mesh of the continuum level model, since the power law relation (see Eq. 4) can be applied to any region of the domain of interest. The CA lattice was independently settled at the beginning of the analysis according to the problem at hand, i.e. the structural architecture of the material element and the physical meaning of the adaptation process. Afterwards, each CA_i was discretized into n_i finite elements by the FE code which automatically set the optimal choice of the refined mesh. At the end of each iteration, when strains and stresses were known at each nodal point of the continuum level FE mesh, a numerical integration procedure based on an 9-point Gaussian quadrature rule (Penninger et al. 2008) allowed for evaluating the strain energy u_{ij} stored inside each finite element j belonging to the i-th CA. Hence, the relevant SED_{i} in each CA_{i} was simply obtained as an average of the energy u_{i} over the volume v_{i} of the cell:

$$SED_{\mathrm{i}} = \frac{u_{\mathrm{i}}}{v_{\mathrm{i}}} = \frac{\sum_{\mathrm{j}=1}^{\mathrm{n_i}} u_{\mathrm{ij}}}{v_{\mathrm{i}}} \tag{6}$$

In this way, new estimates of elastic constants, see Eq. 4, could be incorporated into the new heterogeneous finite element model. With the updated properties, the finite element analyses were repeated and new estimates of the density distribution were calculated at each iteration.

An alternative approach was also adopted, in which the local control rule

$$\frac{dE_{\mathrm{i}}}{dt} = c_{\mathrm{E}}(SED_{\mathrm{i}} - SED^{*}) \tag{7}$$

describes the changes of elastic modulus E_{i} in each cell, in turn converted into changes of the apparent density ρ_{i} by inverting Eq. 4; C_{E} is a constant.

The total strain energy of the discretized bone was approximated as:

$$U = \sum_{\mathrm{i}=1}^{\mathrm{n}} u_{\mathrm{i}} \tag{8}$$

where U_{i} is the strain energy stored in each CA. The total mass M of the structure was given by:

$$M = \sum_{\mathrm{i}=1}^{\mathrm{n}} m_{\mathrm{i}} = \sum_{\mathrm{i}=1}^{\mathrm{n}} m_{\max} x_{\mathrm{i}} \tag{9}$$

where m_{i} is the variable mass, $m_{\max}$ the maximum mass and $x_{\mathrm{i}} = \rho_{\mathrm{i}} / \rho_{\max}$ is the relative mass of the i-th cell, which can range from 0 (no mass) and 1 (maximum density value). Analogously the apparent density was expressed as:

$$\rho_{\mathrm{i}} = \rho_{\max} x_{\mathrm{i}} \tag{10}$$

Following Mullender et al. (1994), Eq. 4 was discretized as:

$$E_{\mathrm{i}} = C\,(\rho_{\mathrm{i}})^{\gamma} = C\,(\rho_{\max} x_{\mathrm{i}})^{\gamma} = [C\,(\rho_{\max})^{\gamma}]\,(x_{\mathrm{i}})^{\gamma} = E_{\max}\,(x_{\mathrm{i}})^{\gamma} \tag{11a}$$

$$y_{\mathrm{i}} = E_{\mathrm{i}}/E_{max} = x_{\mathrm{i}}^{\gamma} \tag{11b}$$

and inverted

$$x_{\mathrm{i}} = \left(\frac{E_i}{E_{max}}\right)^{\frac{1}{\gamma}} = (y_{\mathrm{i}})^{\frac{1}{\gamma}} \tag{12}$$

where E_i is the variable Young's modulus of each CA, E_{max} is the maximum value of the Young's modulus, and $y_i = E_i/E_{max}$ is the relative elastic modulus ($0 \leq y_i \leq 1$). Furthermore, a relationship) between the ultimate stress σ_{ult} and the apparent density of cortical and trabecular bone specimens loaded at low strain rates (Carter and Hayes (1977)) was used:

$$\sigma_{ult} = 68\,\rho^2 = 68\,(\rho_{max}x)^2 \tag{13}$$

Comparison between the Mises stress σ_{Mises} and the ultimate stress σ_{ult} was used to check the structural integrity of the bone tissue by enforcing the inequality $\sigma_{Mises} \leq \sigma_{ult}$.

3 Optimization Procedure

3.1 *Error Signal*

The actual mechanical stimulus SED_i in each *CA* was evaluated via FEM analysis once the external loads, the geometry and the boundary conditions of the structure were defined and a reference mechanical stimulus SED^* was selected. In this study, the initial target SED^*_0 was assumed as the *SED* average of the initial design by means of:

$$SED_0^* = \frac{U_0}{V} \tag{14}$$

where U_0 is total strain energy which is initially stored in the sample volume V. The same value SED^* was assigned to each CA. The error signal in each CA (Huiskes et al. 1987; Beaupré et al. 1990a, b) was calculated by:

$$e_i(t) = \begin{cases} SED_i(t) - (1+l)\cdot SED^* & \text{for } SED_i(t) > (1+l)\cdot SED^* \\ 0 & \text{for } (1-l)\cdot SED^* \leq SED_i(t) \leq (1+l)SED^* \\ SED_i(t) - (1-l)\cdot SED^* & \text{for } SED_i(t) < (1-l)\cdot SED^* \end{cases} \tag{15}$$

where $2l$ is the width of the equilibrium zone. When $e_i(t)=0$ no remodeling occurred and the structure was in equilibrium under the given external loads. For many purposes it is possible to choose $l = 0$ because different values of the equilibrium zone did not seem to be worthwhile to be carried out due to its insignificant influence on the final results as shown in Tovar (2004).

3.2 Local Control Rules

To determine the geometrical or material changes, the remodeling rates, Eqs. 5 or 7, were implemented using a local PID control rule. The time step was arbitrarily set at $\Delta t = 1$. This means that the rate of the simulation process was measured in simulation time and was unrelated to real time (Mullender and Huiskes 1995). The change of apparent density x or, alternatively, of elastic modulus y in each iteration followed from

$$\Delta x_{\mathrm{i}} = x_{\mathrm{i}}(t+\Delta t) - x_{\mathrm{i}}(t) = $$
$$= c_{\mathrm{Px}} \cdot \bar{e}_{\mathrm{i}}(t) + c_{\mathrm{Dx}} \cdot [\bar{e}_{\mathrm{i}}(t) - \bar{e}_{\mathrm{i}}(t-1)] + c_{\mathrm{Ix}} \cdot \sum_{\tau=0}^{t} \bar{e}_{\mathrm{i}}(t-\tau), \quad 1 \le \mathrm{i} \le n \qquad (16)$$

$$\Delta y_i = y_i(t+\Delta t) - y_i(t) = $$
$$= c_{\mathrm{Py}} \cdot \bar{e}_i(t) + c_{\mathrm{Dy}} \cdot [\bar{e}_i(t) - \bar{e}_i(t-1)] + c_{\mathrm{Iy}} \cdot \sum_{\tau=0}^{t} \bar{e}_i(t-\tau), \quad 1 \le \mathrm{i} \le \mathrm{n} \qquad (17)$$

where the coefficients c_{P}, c_{I}, c_{D} are referred to the proportional, integral and derivative gains as indicated by the first subscript, while the second subscript refers to the evolving quantities, namely the mass density and the Young's modulus. The control gains influence the remodeling rate. In particular, the proportional gain determines the direct reaction to the current error. With the derivative control, the change is proportional to the derivative of the error with respect to a unitary change in time $\bar{e}_{\mathrm{i}}(t) - \bar{e}_{\mathrm{i}}(t\text{-}1)$. With the integral control, the change is proportional to the sum of all previous errors.

Moreover, local evolutionary rules were inspired by error-based phenomenological models for bone functional adaptation analysis (Carter 1987; Huiskes et al. 1987; Mullender et al. 1994; Mullender and Huiskes 1995; Huiskes et al. 2000). The local rules made use of the effective error signal defined by:

$$\bar{e}_{\mathrm{i}}(t) = \eta_{\mathrm{i}}(t) \cdot \left[\frac{e_{\mathrm{i}}(t) + \sum_{\mathrm{j}=1}^{\mathrm{N}} \eta_{\mathrm{j}}(t) \cdot e_{\mathrm{j}}(t)}{\mathrm{N}+1} \right] \qquad (18)$$

where $e_{\mathrm{i}}(t)$ is the error signal of the $\mathrm{CA_i}$, $e_{\mathrm{j}}(t)$ is the error signal of the neighbouring $\mathrm{CA_j}$ ($\mathrm{j} = 1, 2, \ldots \mathrm{N}$), N is the number of the neighbouring cells and it can be equal to 0, 4, 8, 12, 24 in two dimensional problems, $\eta_{\mathrm{i}}(t)$ represents the strength of the signal and $\eta_{\mathrm{j}}(t)$ denotes the efficiency of the transmission of the signal from cell "j" to cell "i". In this analysis there is no deterioration in the remodeling signal and therefore $\eta_{\mathrm{i}}(t)=1$ was taken. In addition, also $\eta_{\mathrm{j}}(t)=1$ was valid since values of $0 \le \eta_{\mathrm{j}}(t) < 1$ seemed to be more suitable to describe the effect of microdamage on bone structure behaviour.

The idea of using the influence of CA_i neighbourhoods can be thought of a filtering technique which prevents the appearance of checkerboard patterns and mesh dependency (Patel et al. 2008). Each CA has, in principle, three possibilities to converge and to reach the remodeling equilibrium: (1) the bone is completely resorbed ($\rho = \rho_{min}$); (2) the bone reaches the maximum value of the density ($\rho = \rho_{max}$); or (3) the bone has an apparent density satisfying Eqs. 5 and 16 with $\Delta\rho_i$=0 and, hence, $SED = SED^*$ (Weinans et al. 1992). A saturation density of ρ_{max} was assumed to avoid a density value greater than the bone tissue one (Carter et al. 1989). The local control rules allowed the updating of the relative mass, Eq. 16, or the Young's Modulus, Eq. 17, and the evaluating of the corresponding stress state.

3.3 Finite Element Analysis (FEA)

Stress analyses were performed using Comsol Multiphisics 3.5 software. The single stress analysis came from each step of the iterative process as follows. Strain energy density values from the FEM analysis were the input to specially written algorithms by means of Matlab 7.0.4 code useful to implement Eq. 16. When the evolution of the Young's modulus was considered, Eq. 17 was implemented and the density distribution was obtained by Eq. 12. Updated new density and modulus values were calculated for each CA, becoming the input to the FEM model. The successive iterative calculations were performed.

The iterative process was repeated until the relative change $\Delta U(t)$ in two consecutive iterations, expressed by

$$\frac{\Delta U(t)}{U_0} = \frac{U(t) - U(t-1)}{U_0} \leq \varepsilon \tag{19}$$

was less than a suitably chosen threshold, i.e. $\varepsilon = 0.01\%$. It meant that no more significant changes in the total strain energy stored in the bone tissue occurred. The choice of the strain energy as parameter of the stopping rule was motivated by checking the global stiffness of the structure. In Fig. 1 a diagram block shows the adaptive remodeling procedure which combines changes in modulus and geometry of the investigated domain and changes in strain energy based on FEM analysis.

3.4 Cost Indices

Maximizing stiffness means minimizing the strain energy and this goal can be accomplished by defining the following cost indices:

$$J_1 = (1 - \omega) \cdot \frac{M}{M_0} + \omega \cdot \frac{U}{U_0} \tag{20}$$

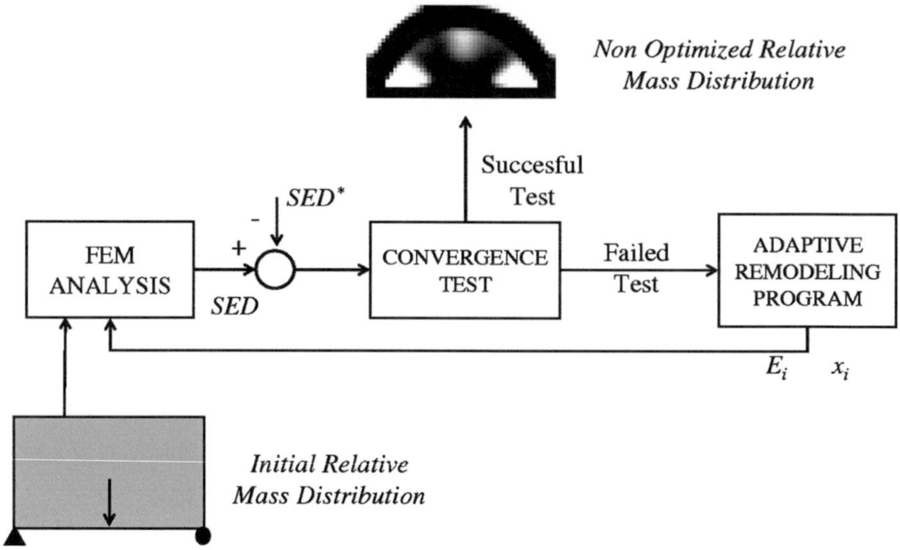

Fig. 1 Control modulus for adaptive remodeling process

or alternatively (Andreaus et al. 2011):

$$J_2 = \frac{M}{M_0} \cdot \frac{U}{U_0} \tag{21}$$

where U_0 and M_0 are respectively the total strain energy and the mass of the domain to be optimized when $x_i = x_0$ (for i = 1,2,...n). In the cost index J_1 the relative influence of the mass and the energy terms is balanced by the weight parameter ω with values between 0 and 1.

In the optimization theory the weight ω is usually dependent on the faced problem. In this case, during the optimization process ω can attain any value in the admissible range. In fact, it is hypothesized that functional adaptation redistributes mass or energy in an optimal way with respect to its own physiological requirements.

The PID gains (c_P, c_I, c_D) in Eq. 16 or 17, the target (*SED**) in the error signal ($e_i(t) = SED_i - SED^*$), and the multiplier ω of normalized total mass and energy in Eq. 20 were optimized by minimizing the cost indices J_1, Eq. 20, or J_2, Eq. 21, as explained in the next Subsection.

3.5 Non Linear Constrained Minimization Problem

The global, multi-parameter optimization problem, in which stiffness and mass are conflicting objectives, was expressed as the evaluation of an optimal choice for the

PID control gains c_P, c_I, c_D, the target *SED** and the weight parameter ω in order to minimize the alternative cost indices, Eqs. 20 and 21:

$$MinJ_1 = \underset{(c_P,c_I,c_D,SED^*,\omega)}{Min} \left[(1-\omega)\frac{M(c_P,c_I,c_D,SED^*)}{M_0} + \omega\frac{U(c_P,c_I,c_D,SED^*)}{U_0} \right] \tag{22}$$

or

$$MinJ_2 = \underset{(c_P,c_I,c_D,SED^*)}{Min} \left[\frac{M(c_P,c_I,c_D,SED^*)}{M_0} \cdot \frac{U(c_P,c_I,c_D,SED^*)}{U_0} \right] \tag{23}$$

The cost indices J_1 and J_2 explicitly depend upon total mass M and energy U, which in turn depend on the distributions of the mass density (see Eq.16) and of the elastic modulus (see Eq. 11a), respectively. In more details, Eq. 16 depends on the PID gains and on the effective error signal, see Eq. 18 with $\eta(t)=1$, and hence, on the target in the error signal $e_i(t) = SED_i - SED^*$ at each iteration. Unfortunately, the above mentioned relationships are not definitely known in closed form, and only FEM analysis and optimization programming can assess them as a result of an evolution process.

The optimization variables were constrained by the following inequalities:

$$c_P \geq 0, \quad c_I \geq, \; 0, \quad c_D \geq 0, \quad SED^* \geq 0, \quad 0 \leq \omega \leq 1 \tag{24}$$

Problems (22) and (23) in addition to conditions (24) are non linear constrained optimization problems. There is no analytical solution because closed-form expressions of the total mass and of the total energy are not available. Therefore a numerical solution is essential. To speed up the convergence and to avoid local minima, it was convenient to constrain more strictly the control parameters and the target. The choice of the constraint limits was based on a significant number of trial-and-error tests according to the control procedure illustrated by the flow-chart of Fig. 1. *A posteriori* evaluating of the cost indices (20) or (21) allowed the confinement of the admissible set of values for the parameters to be optimized. The constraints considered for the numerical solution of problems (22) and (23) were:

$$\begin{aligned} &c_P^{min} \leq c_P \leq c_P^{max} \\ &c_I^{min} \leq c_I \leq c_I^{max} \\ &c_D^{min} \leq c_D \leq c_D^{max} \\ &SED^{*\,min} \leq SED^* \leq SED^{*\,max} \\ &0 \leq \omega \leq 1 \end{aligned} \tag{25}$$

and, along with the characteristics of the objective function, admitted the application of the *Weierstrass* theorem, guaranteeing the existence of the optimal solution. Furthermore, the safety condition to be respected was established as

$$\sigma_{Mises} < \sigma_{ult} \tag{26}$$

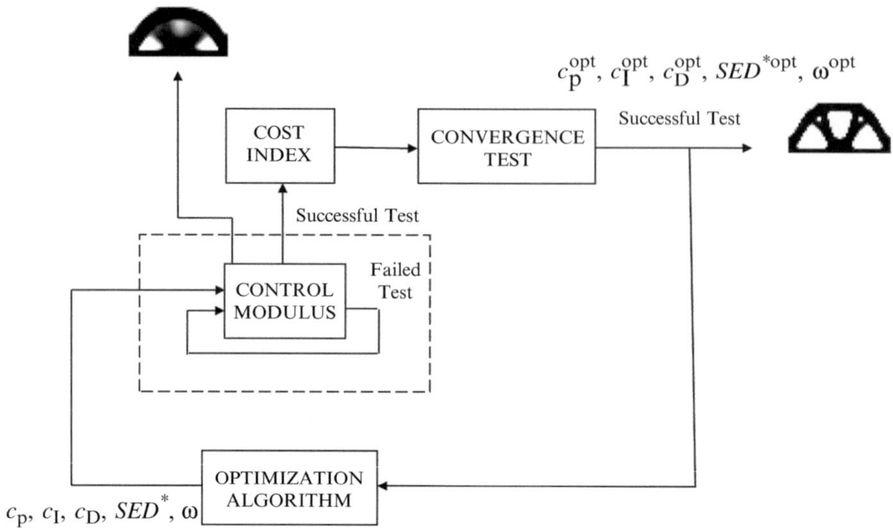

Fig. 2 Block diagram of the proposed optimal control procedure

In order to avoid the violation of this condition, a penalty term *P* was added to the chosen cost index (Eq. 20 or 21):

$$J(c_P, c_I, c_D, SED^*, \omega) + P(c_P, c_I, c_D, SED^*, \omega) \tag{27}$$

In general, the penalty term is chosen in different forms. In this study, it is represented by the maximum of the absolute value attained at each step by the left-hand side of Eq. 26 for the cellular automata in which the safety condition was violated:

$$P = B \cdot \left\{ \underset{i=1,n}{Max} \left[\left| 1 - (\sigma_{Mises}/\sigma_{ult})_i \right| \right] \right\} \tag{28}$$

when $[1\text{-}(\sigma_{Mises}/\sigma_{ult})_i] \leq 0$, where *B* is the penalization weight.

In Fig. 2 the block diagram relative to the proposed procedure is displayed. It is important to note that the control scheme of Fig. 1 is a sub-module of the optimization procedure which can be explained as follows: an initial condition for the parameters to be optimized was assigned; for instance, the initial value SED^*_0 of the target SED^* was assumed as the average value of the *SED* corresponding to the Young's modulus E_0 evaluated with initial relative mass x_0. The FEA and the remodeling algorithms returned a value for the chosen cost indices. Successively, the optimization procedure suitably modified the set of parameters (c_P, c_I, c_D, SED^*,ω). Again, the FEM and remodeling algorithms returned a new value of the cost indices based on the new set of parameters, and cycling did continue till convergence test was successful, that is the increment of the total energy became negligible with respect to a fixed tolerance, Eq. 19. The optimization procedure stopped the iterative cycle when the first order optimality conditions were satisfied to the specified tolerance of 1×10^{-6}

In Fig. 2, the optimization algorithm is characterized by the function *fmincon* of ©Matlab. It is implemented by means of the sequential quadratic programming (SQP) method (Fletcher 1987) with the aim to finding a constrained minimum of a function of several variables. At each iteration, the function solves a quadratic programming subproblem (Gill et al. 1981) in which a positive definite quasi-Newton approximation of the Hessian of the Lagrangian function is updated using the Broyden–Fletcher–Goldfarb–Shanno (BFGS) method (Battiti 1992).

4 Two-Dimensional Model

The method proposed in the previous Sects. 2 and 3 is tested to optimize the topology of a Michell-type two-dimensional structure as a continuum design domain. This sample works out to predict optimal configurations of the trabecular bone internal architecture. The design domain is characterized by a rectangular area of 50×25 mm^2 and by a thickness of 1 mm, as applied earlier by Tovar (2004) and Tovar et al. (2007). It is divided into 1,250 identical cells, see Fig. 3.

Each cell is discretized by quadratic triangular Lagrange elements with six nodes at the corners and side midpoints. One of the lower corners is restrained from vertical and horizontal displacement, while the displacement of the opposite lower

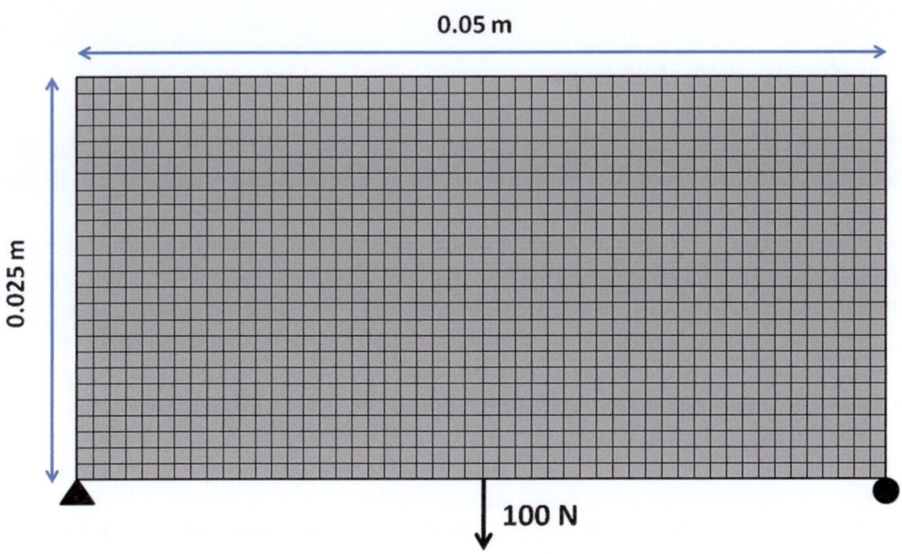

Fig. 3 Boundary and loading conditions applied to the Michell-type structure

corner is constrained only in the vertical direction. A vertical load of 100 N is applied in the middle of the lower edge.

A uniform initial density distribution of $\rho_0 = \rho_{max} = 1{,}740\ \mathrm{Kg/m^3}$ was assumed. The initial Young's modulus, equal to $E_0 = E_{max} = 17$ GPa, represents the stiffness of femoral cortex longitudinally loaded in compression (Weinans et al. 1992). The relation between Young's modulus E and the apparent density ρ is taken by Mullender and Huiskes (1995) with $\gamma = 3$ and $C = 3.227\ \mathrm{GPa/(Kg/m^3)^3}$ in Eq. 11a. As a reasonable approximation, it is assumed that the Poisson's ratio $\nu = 0.3$ is constant and does not change when the local bone apparent density changes. The validity of Eq. 4 and the assumption of constant value of Poisson's ratio were addressed by Carter et al. (1989). The minimum density value is $\rho_{min} = 0.01 \cdot \rho_{max}$, representing complete resorption of a cellular automaton.

The initial average value of the strain energy density under the assigned load was selected as initial target. In order to calculate the mean *SED* per unit area, homogeneous mass density and associated material properties were first assigned to all elements. The strain energy density SED_i in each CA was calculated by Eq. 6. The initial target SED^*_0 was approximately evaluated by:

$$SED_0^* = \frac{U_0}{A} = 2.5\ \left(\mathrm{J/m^2}\right) \tag{29}$$

where U_0 is the initially stored total strain energy, see Eq. 8, and A is the domain area.

Two comparative analyses were performed by taking the rate of change of the relative mass x_i and the Young's modulus E_i as free variables and by expressing them as a function of the effective error signal $\bar{e}_i(t)$ through a PID controller.

5 Results and Discussion

Evolutions of total optimized mass and energy are displayed in Figs. 4 (see Eq. 16) and 6 (see Eq. 17), assuming the initial relative mass uniformly equal to 1.0. In the same initial conditions the corresponding evolutions of the cost indices J_1 and J_2 are shown in Figs. 5, 7 and evaluated with the optimized values of the relevant parameters. The optimization process required 98 iterations of the optimization procedure.

In more details, under the implemented relative mass rule the nested control procedure for the *Michell-type Structure*, discretized into 1,250 cellular automata, required 60 iterations to converge to an optimal state (see Figs. 4 and 5).

In Tables 1 and 2 the initial conditions, the constraint limits of the parameters at the beginning of the optimization process and the optimal parameters values at the end of the optimization process are summarized.

The optimal value of the weight ω is 0.23, meaning that the mass term was weighted more significantly with respect to the sample worked out in (Andreaus et al. 2012). Indeed, Fig. 4 shows a halved final value of the bone mass if compared to the initial one.

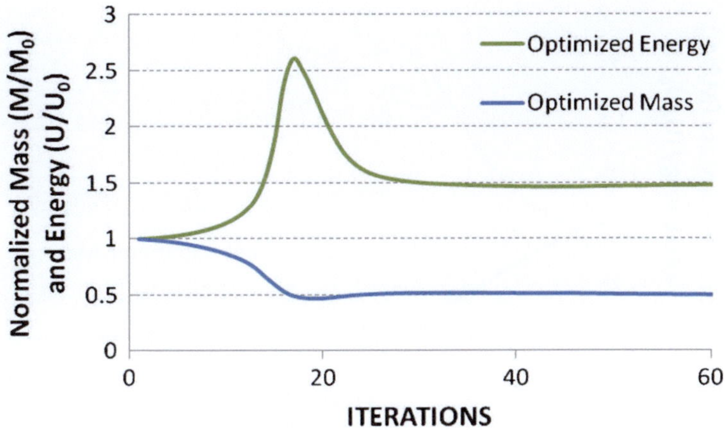

Fig. 4 Evolution of total optimized mass and energy through the implemented relative mass rule The initial relative mass x_0 is uniformly assumed equal to 1.0

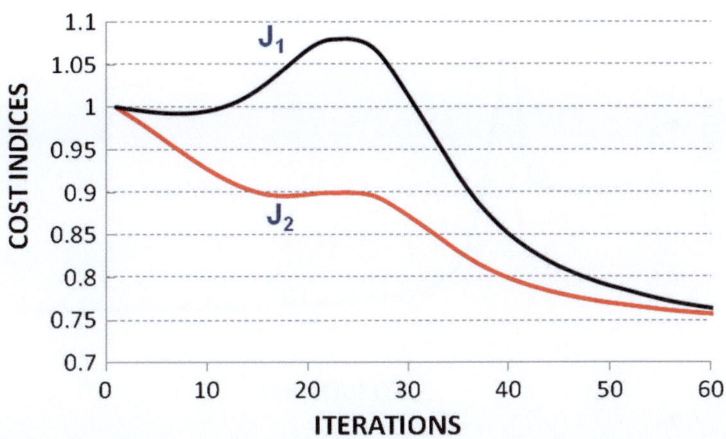

Fig. 5 Evolution of the cost indices J_1 and J_2 minimized by the optimal parameters. The relative mass rule is implemented

When the Young's modulus rule was implemented, the nested control procedure with the initial condition of $x_0 = 1$ required 80 iterations to converge to an optimal state (see Figs. 6 and 7).

Tables 2 and 4 show the initial conditions, the constraint limits of the parameters at the beginning of the optimization process and the optimal parameters values at the end of the optimization process are summarized for the corresponding evolution rule.

The optimal value of the weight ω is 0.24, and also in this case the mass term was weighted more significantly, showing a gain of 50% with respect to the initial value (see Fig. 6). For both cases, the cost index J_2 revealed the best behaviour in the plots.

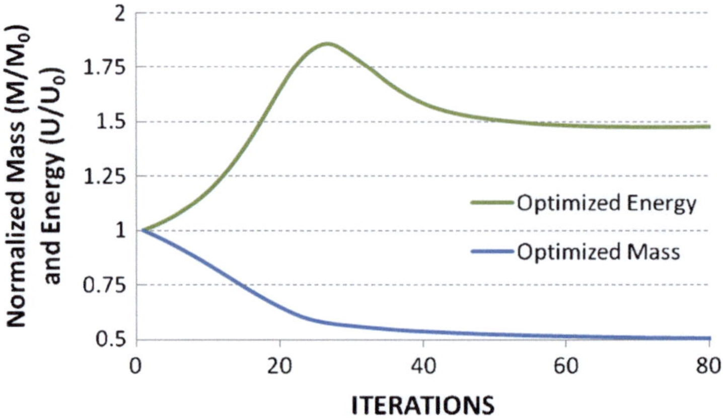

Fig. 6 Evolution of total optimized mass and energy through the implemented Young's modulus rule. The initial relative mass x_0 is uniformly assumed equal to 1.0

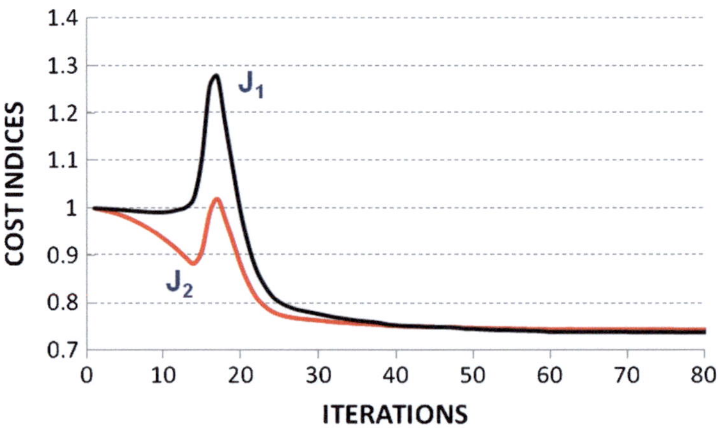

Fig. 7 Evolution of the cost indices J_1 and J_2 minimized by the optimal parameters. The Young's modulus rule is implemented

Table 1 Initial conditions and constraints limits for the parameters of the control and optimization procedure. The relative mass evolution rule is applied

Parameter	x	c_P	c_I	c_D	SED * [J/m^2]	ω
t_0	1	0.0525	0.00275	0.00275	2.5	0.5
min	0	0.005	5×10^{-4}	5×10^{-4}	1	0
max	1	0.1	0.05	0.05	5	1

Table 2 Optimal parameter values when the relative mass evolution rule is implemented

Cost indices	c_P^{opt}	c_I^{opt}	c_D^{opt}	SED^{*opt} [J/m^2]	ω^{opt}
J_1^{opt}	0.02	0.001	0.001	2.18	0.23
J_2^{opt}	0.01	0.001	0.001	2	

Table 3 Initial conditions and constraints limits for the parameters of the control and optimization procedure. The Young's modulus evolution rule is applied

Parameter	x	c_P [(Pa·m)/J^2)]	c_I [(Pa·m)/J^2)]	c_D [(Pa·m)/J^2)]	SED *[J/m^2]	ω
t_0	1	$2.52\times10^{+8}$	$2.52\times10^{+8}$	$2.52\times10^{+8}$	2.5	0.5
min	0	$5.5\times10^{+6}$	$5.5\times10^{+6}$	$5.5\times10^{+6}$	1	0
max	1	$5\times10^{+8}$	$5\times10^{+8}$	$5\times10^{+8}$	5	1

Table 4 Optimal parameters when the Young's modulus evolution rule is implemented

Cost indices	c_P^{opt}[(Pa·m)/J^2]	c_I^{opt}[(Pa·m)/J^2)]	c_D^{opt}[(Pa·m)/J^2)]	SED^{*opt} [J/m^2]	ω^{opt}
J_1^{opt}	$10\times10^{+7}$	$10\times10^{+7}$	$10\times10^{+7}$	2.18	0.24
J_2^{opt}	$15\times10^{+7}$	$10\times10^{+7}$	$8\times10^{+7}$	2.18	

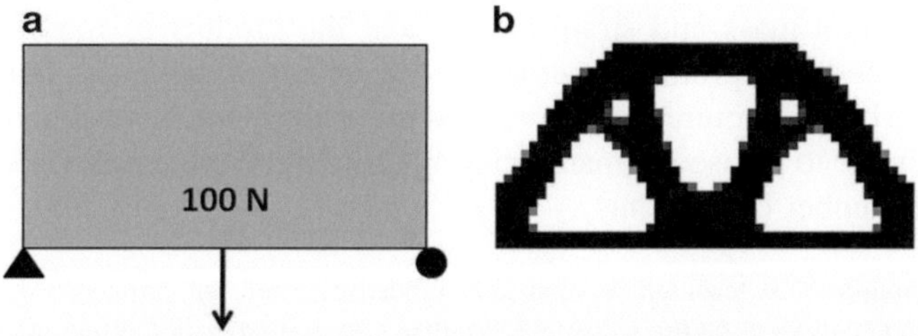

Fig. 8 Initial relative mass distribution (**a**) and final relative optimal mass distribution (**b**). The bone structure is initially considered homogeneous and isotropic ($x_0 = 1$)

As shown in Tables 2 and 4, the optimal target is equal to 2.18 J/m^2 for both implemented relative mass and Young's modulus rules and it is lower than the initial value.

To conclude, in Fig. 8 the configurations of the initial and final relative mass distributions of the trabecular structure are shown. The bone structure was initially considered homogeneous and isotropic ($x_0 = 1.0$). At the end of the optimization process it became heterogeneous consistently in applied boundary and loading conditions.

In general, for a refined discretized domain, the geometric characteristics of the solution depend only on the applied loads, the value of the target to reach and the maximum attainable elastic modulus or maximum apparent density.

6 Conclusions

This work achieved the goal of modeling a two-dimensional trabecular structure having maximum stiffness and minimum weight in an efficient computational way. An implemented algorithm allowed the interaction between evolution rules of the considered material and structural analysis of the model under given loads. The novelty of the proposed method is three-fold: first, the use of alternative evolution rules, expressed in Eqs. 16 and 17, and the implementation of two consecutive control and optimization procedures; second, the definition of alternative cost indices J_1 and J_2; third, the structural application to trabecular bone tissue. In more details, a PID control was assumed with respect to both evolution rules of mass and stiffness. The process of material remodeling comprised two stages: the first stage consisted of a control procedure (trial and error) which provided refined initial values of the involved parameters (control gains, target and weight ω) to start the subsequent optimization procedure, as second stage, which – in turn – led to the optimal solution of the minimization problems. Moreover, to the knowledge of the authors, the second cost index J_2 defined as the product of mass and energy was newly introduced.

The automatic code was characterized by rapid convergence at low computational cost in terms of computation time and memory occupation. These advantages allowed for comparing the above mentioned cost indices, the linear combination of suitably weighted mass and strain energy and the product of mass and energy, namely. A good compromise seemed to be achieved with the latter one. To conclude, the bone structure was discretized in 1,250 cell elements but the proposed model is suitable to be used in many other fields of innovative materials discretized in whatever number of elements.

Acknowledgments This research was partially funded by Sapienza University of Rome under *Progetto di Ateneo 2004* grant no. C26A044385, *Progetto di Ateneo 2005* grant no. C26A059503 and *Progetto di Università 2008* grant no. C26A08E7B3.

References

Andreaus U, Colloca M, Iacoviello D, Pignataro M (2011) Optimal-tuning PID control of adaptive materials for structural efficiency. Struct Multidiscip Optim 43:43–59

Andreaus U, Colloca M, Iacoviello D (2012) An optimal control procedure for bone adaptation under mechanical stimulus. Control Eng Pract 20:575–583

Bagge M (2000) A model of bone adaptation as an optimization process. J Biomech 33:1349–1357

Battiti R (1992) First and second order methods for learning: between steepest descent and Newton's method. Neural Comput 4:141–166

Beaupré GS, Orr TE, Carter DR (1990a) An approach for time-dependent bone modeling and remodeling-theoretical development. J Orthop Res 8:651–661

Beaupré GS, Orr TE, Carter DR (1990b) An approach for time-dependent bone modeling and remodeling-application: a preliminary remodeling simulation. J Orthop Res 8:662–670

Bourgery JM (1832) Traité complet de l'anatomie de l'homme. Delaunay, Paris

Carter DR (1984) Mechanical loading histories and cortical bone remodeling. Calcif Tissue Int 36:519–524

Carter DR (1987) Mechanical loading history in skeletal biology. J Biomech 20:1095–1105

Carter DR, Hayes WC (1977) The compressive behaviour of bone as a two-phase porous structure. J Bone Joint Surg Am 59:954–962

Carter DR, Fyhrie DP, Whalen RT (1987) Trabecular bone density and loading history: regulation of connective tissue biology by mechanical energy. J Biomech 20:785–794

Carter DR, Orr TE, Fyhrie DP (1989) Relationships between loading history and femoral cancellous bone architecture. J Biomech 22:231–244

Cowin SC, Doty SB (2007) Tissue mechanics. Springer, New York

Cowin SC, Hegedus DH (1976) Bone remodeling I: a theory of adaptive elasticity. J Elast 6:313–326
Currey JD (1988) The effect of porosity and mineral content on the Young's modulus of elasticity of compact bone. J Biomech 21:131–139
Curry J (1984) The mechanical adaptations of bones. Princeton University Press, Princeton
Fletcher R (1987) Practical methods of optimization, 2nd edn. John Wiley and Sons, Chichester
Frost HM (1983) A determinant of bone architecture. The minimum effective strain. Clin Orthop Relat Res 175:286–292
Fyhrie DP, Carter DR (1986) A unifying principle relating stress to trabecular bone morphology. J Orthop Res 4:304–317
Gill PE, Murray W, Wright MH (1981) Practical optimization. Academic, London
Huiskes R, Weinans H, Grootenboer J, Dalstra M, Fudala M, Slooff TJ (1987) Adaptive bone remodeling theory applied to prosthetic-design analysis. J Biomech 20:1135–1150
Huiskes R, Ruimerman R, van Lenthe GH, Janssen JD (2000) Effects of mechanical forces on maintenance and adaptation of form in trabecular bone. Nature 405:704–706
Jacobs CR, Simo JC, Beaupré GS, Carter DR (1997) Adaptive bone remodeling incorporating simultaneous density and anisotropy considerations. J Biomech 30:603–613
Lekszycki T (2005) Functional adaptation of bone as an optimal control problem. J Theor Appl Mech 43:555–574
Lennartson B, Kristiansson B (2006) Evaluation and tuning of robust PID controllers. IET Control Theor Appl 3:294–302
Liu GP, Daley S (1999) Optimal-tuning PID controller design in a frequency domain with application to a rotary hydraulic system. Control Eng Pract 7:821–830
Liu GP, Daley S (2001) Optimal-tuning PID control for industrial system. Control Eng Pract 9:1185–1194
Mullender MG, Huiskes R (1995) Proposal for the regulatory mechanism of Wolff's law. J Orthop Res 13:503–512
Mullender MG, Huiskes R, Weinans H (1994) A physiological approach to the simulation of bone remodeling as a self-organizational control process. J Biomech 27:1389–1394
Patel NM, Tillotson D, Renaud JE, Tovar A, Izui K (2008) Comparative study of topology optimization techniques. AIAA J 46:1963–1975
Penninger CL, Patel NM, Niebur GL, Tovar A, Renaud JE (2008) A fully anisotropic hierarchical hybrid cellular automaton algorithm to simulate bone remodeling. Mech Res Commun 35:32–42
Roesler H (1987) The history of some fundamental concepts in bone biomechanics. J Biomech 20:1025–1034
Roux W (1881) Der ziichtende kampf der teile, oder die 'teilauslese' im organismus (theorie der 'finktionellen unpassung'). Wilhelm Engelmann, Leipzig
Rubin CT (1984) Skeletal strain and the functional significance of bone architecture. Calcif Tissue Int 36:11–18
Rubin CT, Lanyon LE (1987) Osteo-regulatory nature of mechanical stimuli: function as a determinant for adaptive remodeling in bone. J Orthop Res 5:300–310
Subbarayan G, Bartel DL (1989) Fully stressed and minimum mass structures in bone remodeling. In: Friedman MH (ed) Biomechanics symposium, vol 98. American Society of Mechanical Engineers, New York, pp 173–176
Tovar A (2004) Bone remodeling as a hybrid cellular automaton optimization process. Dissertation, University of Notre Dame
Tovar A, Patel NM, Niebur GL, Sen M, Renaud JE (2006) Topology optimization using a hybrid cellular automaton method with local control rules. ASME J Mech Des 128:1205–1216
Tovar A, Patel NM, Kaushik AK, Renaud JE (2007) Optimality conditions of the hybrid cellular automata for structural optimization. AIAA J 45:673–683
Turner CH (1991) Homeostatic control of bone structure: an application of feedback theory. Bone 12:203–217
Weinans H, Huiskes R, Grootenboer HJ (1992) The behaviour of adaptive bone-remodeling simulations models. J Biomech 25:1425–1441
Wolff J (1892) Das Gesetz der transformation der knochem. Hirschwald, Berlin

The Kinematics of the Hip Joint with Femoroacetabular Impingement May Be Affected by the Thickness of the Articular Cartilage

Radhakrishna Suppanee, Prudence Wong, Ibrahim Esat, Mahmoud Chizari, Karthig Rajakulendran, Nikolaos V. Bardakos, and Richard E. Field

Abstract The abnormalities in the shape and orientation of the femoral head and neck or the acetabulum are important morphological characteristics of femoroacetabular impingement (FAI). Concerns exist about the effect of damaged cartilage on the kinematics of the affected hip joint. The current study is aiming to track the motion of a femur bearing a cam deformity, with healthy or damaged articular cartilage. This may prove useful in understanding the changes occurring in a hip joint with cam-type FAI, as arthritis develops and progresses. A three-dimensional (3D) model of the left hip joint of a male patient diagnosed with FAI was obtained from pre-operative Computerised Tomography (CT) data using density segmentation techniques in Mimics 13.1 (Materialise NV). The kinematics of FAI was analysed in Abaqus 6.9 (Simulia Dassault Systems) using a finite element method. The translation and rotation parameters were defined in a single step for each one of three cases: healthy cartilage, 2 mm (one-sided thinning) and 4 mm (two-sided thinning) worn-out articular cartilage. As the acetabulum and femur came into contact, the penetrations were detected and the contact constraints were applied according to the penalty constraint enforcement method. The results of the analysis showed that thinning of the cartilage at the hip joint adversely affects impingement, as range of motion was decreased with progressive thinning of the articular cartilage.

R. Suppanee • P. Wong • I. Esat
School of Engineering and Design, Brunel University West London, Uxbridge, UK

M. Chizari (✉)
Orthopaedic Research and Learning Centre, School of Engineering and Design, Brunel University West London, Uxbridge UB8 3PH, UK
e-mail: mahmoudchizari@yahoo.com

K. Rajakulendran • N.V. Bardakos • R.E. Field
The South West London Elective Orthopaedic Centre, Epsom, UK

D. Iacoviello and U. Andreaus (eds.), *Biomedical Imaging and Computational Modeling in Biomechanics*, Lecture Notes in Computational Vision and Biomechanics 4, DOI 10.1007/978-94-007-4270-3_3,

Keywords Thinning of the articular cartilage • Kinematic analysis • Hip impingement • Finite element modeling

1 Introduction

One of the most common causes of enfeebling pain in the general population has been hip osteoarthritis (OA) (Ingvarsson 2000). The two main categories of hip OA are primary, also known as idiopathic, and secondary hip OA from a known cause e.g. Perthes disease, developmental dysplasia of the hip, post-traumatic, osteonecrosis (Dagenais et al. 2009). In fact, the aetiology of osteoarthritis of the hip has long been considered secondary to congenital or developmental deformities or primary whilst presuming some underlying abnormality of the articular cartilage (Ganz et al. 2008). However, recent information supports the hypothesis that the so-called primary osteoarthritis is also secondary to subtle developmental abnormalities. The mechanism in these cases is femoroacetabular impingement (FAI) rather than excessive contact stress (Ganz et al. 2008). The distinction here lies in the novel concept that more subtle deformities, often unrecognized in the past, can damage the labrochondral junction, ultimately leading to hip OA. Two primary types of impingement have been distinguished based on the origin and the mechanism of impingement, commonly referred to as pincer and cam (Tannast et al. 2008).

In this study, we focused on cam impingement. During cam impingement, the labrum remains uninvolved over a rather long period (Ganz et al. 2008). In fact, avulsion of the acetabular articular cartilage from the labrum and then delamination from the subchondral bone occurs (Ganz et al. 2008). Cam impingement is considered more destructive than pincer (Ganz et al. 2008). Left untreated, it may cause the hip to fail toward anterosuperior osteoarthritis (Ganz et al. 2008). Figure 1 shows an arthroscopic intra-operative photograph of a hip with cam-type femoroacetabular impingement.

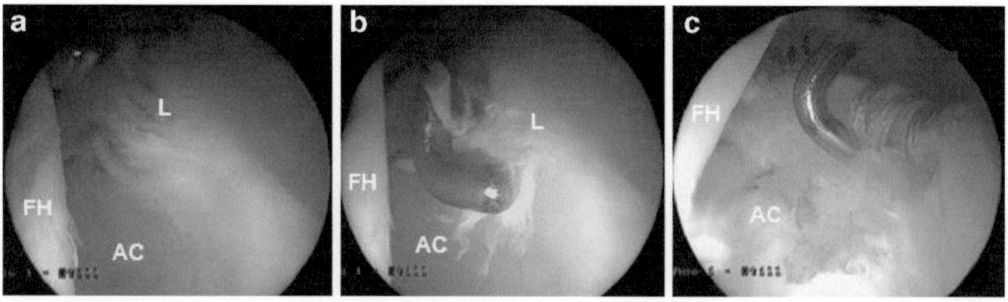

Fig. 1 (**a**) Arthroscopic intra-operative photograph of a hip with cam-type femoroacetabular impingement, showing severe delamination of the articular cartilage of the acetabulum at the anterosuperior labrochondral junction, and (**b**) the same lesion pushed with an arthroscopic probe. (**c**) The loose cartilaginous flap was debrided and the underlying subchondral bone was microfractured. The overlying labrum was found to be detached and was repaired with use of two suture anchors. ac, acetabular cartilage; fh, femoral head; l, labral

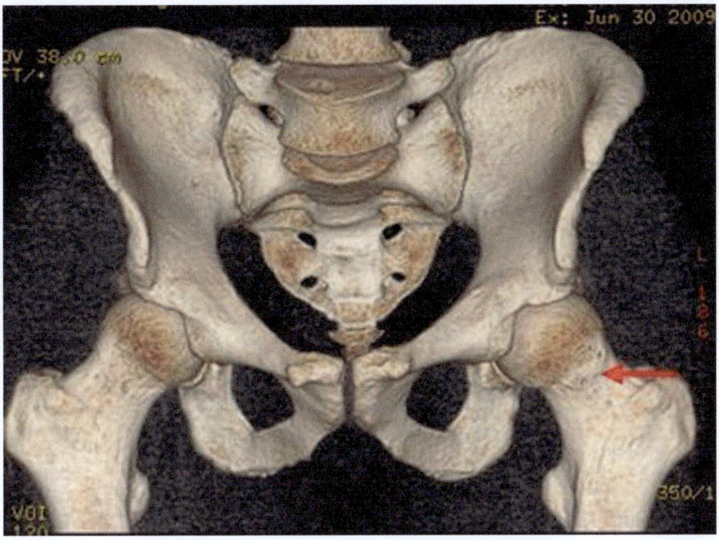

Fig. 2 A volume rendering of the hip joints and pelvis of the male patient with an osseous bump (*arrow*) at the anterosuperior head-neck junction of the left hip which causes cam impingement on the rim of the acetabulum

Figure 2 shows the volume rendering of the hip joints and pelvis of a male patient who underwent a Computed Tomography (CT) scan. The volume rendering of the hip joint was viewed using Sectra Image Viewer software. It was observed that there was a subtle deformity at the anterosuperior head-neck junction of the patient's hip that gave rise to cam impingement. The impingement was found to be located at the anterosuperior rim of the acetabulum. The right hip was comparatively normal.

The patient also underwent a magnetic resonance (MR) arthrogram of the left hip joint which showed a partial thickness tear of the articular surface of the anterior cartilaginous labrum adjacent to the bony labrum (Fig. 3). However, the arthrogram revealed that there were no significant abnormalities in the articular cartilage and no fibrocystic change was seen in the head of the femur or acetabulum.

Concerns exist about the effect of damaged cartilage on the kinematics of the affected hip joint. The current study is aiming to track the motion of a femur bearing a cam deformity, with healthy or damaged articular cartilage. This may prove useful in understanding the changes occurring in a hip joint with cam-type FAI, as arthritis develops and progresses.

It was hypothesised that: the impingement would occur at the anterosuperior rim of the acetabulum, the risk of impingement would increase by thinning of the articular cartilage, and flexion, adduction and internal rotation would decrease compared to normal hips and with progressive thinning of the articular cartilage of the femoral head and/or the acetabulum.

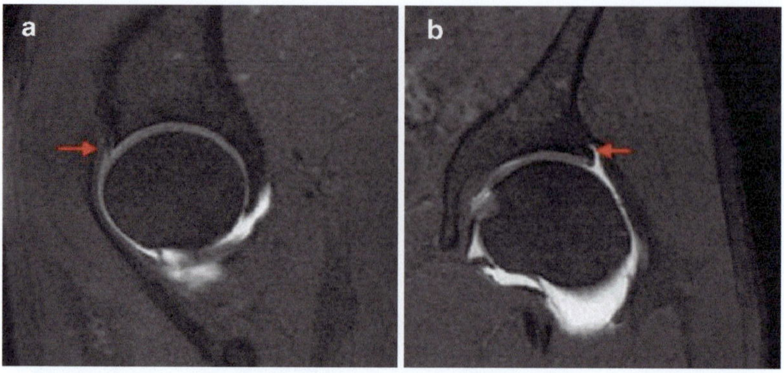

Fig. 3 (**a**) A sagittal image of the partial thickness tear (*arrow*) of the articular surface of the anterior cartilaginous labrum adjacent to the bony labrum. (**b**) A coronal image of the same (*arrow*)

2 Methodology

The following objectives were set up: acquisition of a three-dimensional (3-D) anatomical geometry of the femur bones of a male patient from a CT and MRI data using Mimics software 13.1 (Materialise NV), construction of a finite element model in Abaqus 6.9-1 (Simulia Dassault Systems) using the acquired 3D geometrical hip joint, definition of the translation and rotation parameters for three cases:

- Healthy cartilage with no sign of thinning on the articular cartilage of both the femur head and acetabulum at the contact interface
- 2 mm thinning either at the femur head or at the acetabulum where the contact occurs
- 4 mm thinning on both femur head and acetabulum at the contact region

The above assumption was made to simulate the physiological kinematic values that cause cam impingement for the above mentioned cases.

Mimics software was used to read the Digital Imaging and Communications in Medicine (DICOM) files of the subject patient. Segmentation and editing tools allow the manipulation of data to select bone and soft tissue. After generating the 3-D model it was surface meshed using triangular elements. The surface mesh was then exported to Abaqus and converted to a volume mesh (tetrahedral elements) prior to post processing.

2.1 *Methods for the 3D Modelling*

The anthropometrical data of a 41 years old male patient with a height of 195 cm and weight of 106.5 kg at the time of surgery with bone abnormalities indicative of

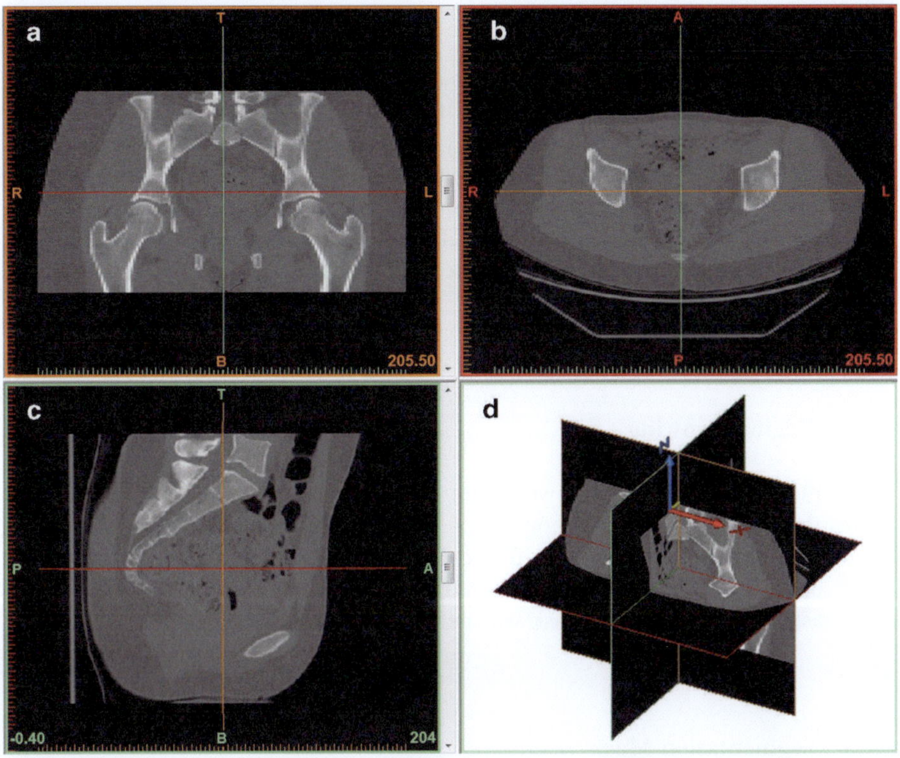

Fig. 4 (**a**) A single slice of a gray value image (DICOM format) of the patient's hip joint in the coronal plane, (**b**) axial plane, (**c**) sagittal plane. (**d**) Multiplanar rendering of the gray slices of the hip joint in the coronal, sagittal and axial views

cam-type deformity at the anterosuperior head-neck junction was acquired using GE Medical systems/Light speed VCT computed tomography scan. The patient was scanned from the pelvic girdle to the distal end of the femur in the supine position. The scans were made up of 1027 cross-sectional slices with a slice distance of 0.625 mm and a field of view (FOV) of 380 mm. The DICOM images generated during the CT scan were processed using Mimics software to obtain the primary 3D model using density segmentation techniques. 1027 image slices in the DICOM format were selected and automatically imported to Mimics.

A pixel size of 0.7422 mm was automatically computed accounting the present image resolution which was 512 × 512 pixels. In fact, the pixel size and the resolution of the image ensure the model generated is dimensionally coherent during the segmentation process (Mimics Materialise 13.1). The image slices were then stacked and converted to be displayed in the coronal, sagittal and axial views. However, the orientation of each view was defined before continuing (Fig. 4). Once the 3D model of the hip joint was obtained, the mesh discretisation process was carried out in Mimics and subsequently material constitutive, loads and boundary conditions, and kinematics constraints were defined in Abaqus software.

2.1.1 Thresholding

Thresholding based on the Hounsfield scale was used to separate both the cortical bone of the femur and acetabulum from the surrounding muscles and tissues. CT images are a pixel map of the linear X-ray attenuation coefficient of tissue (Mimics Materialise 13.1). Mimics software allows changing the predefined threshold to the required threshold by choosing compact bone of an adult from a drop down menu. Therefore, a lower limit of 662 Hounsfield units (HU) and upper limit of 1652 HU were defined in order to select only the cortical bones from the surrounding tissues (Fig. 5).

2.1.2 Region Growing

The femur and the pelvic girdle were then selected separately and assigned different masks through region growing operation. A point was clicked on the area of the femur and the program then started to calculate the new segmentation. That is all points on the area of the femur connected to the marked point were used to form a new mask. Similarly, a new mask was defined for the pelvic girdle (Fig. 6).

2.1.3 Segmentation Mask

The region growing has split the segmentation into separate entities. This enables to generate separate geometrical files and subsequently the 3-D model. Noisy pixels and artefacts were eliminated manually. Cavity filling operation was performed in order to produce density masks of both the femur and the pelvic girdle. A 3-D computation was then carried out on both the density masks (Fig. 6). Both the 3-D models were wrapped and conservatively smoothened using a smooth factor of 0.4 in order to have a good surface topography without affecting the anthropometrical data of the patient.

2.2 *Generation of Surface Mesh*

The 3-D models were separately divided into triangular elements and nodes (surface meshing) in Mimics remesher module. A quality threshold of 0.3 was defined for the triangular elements. The height/base parameter was used to measure the ratio between the height and the base of a triangle.

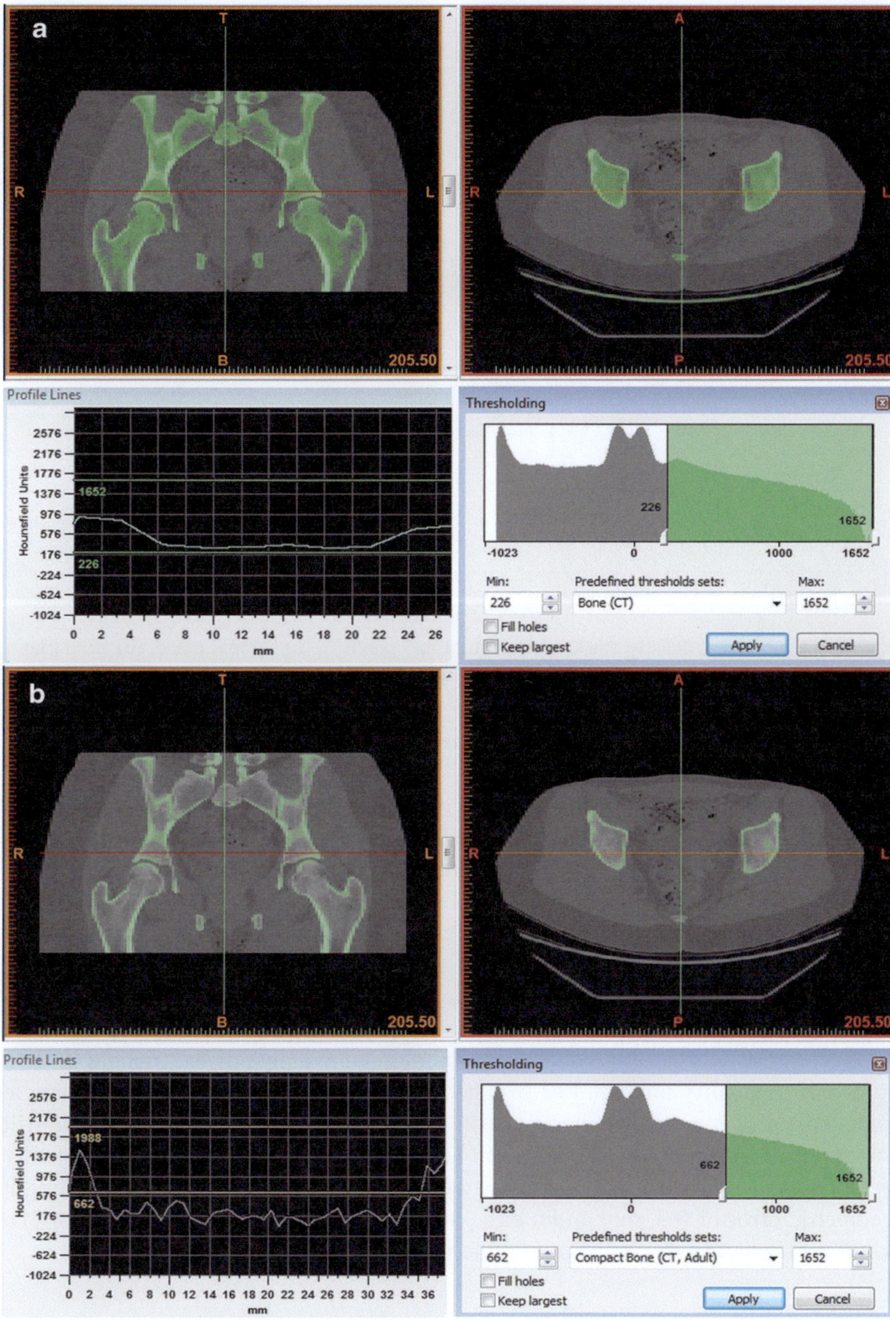

Fig. 5 (**a**) The Hounsfield threshold of the bone with the corresponding intensity profile (a lower limit of 226 HU and upper limit of 1652 HU) and histogram. (**b**) Adjustment of the Hounsfield threshold for the selection of the compact bone only. The intensity profile and histogram was adjusted to a lower limit of 662 HU and upper limit of 1652 HU

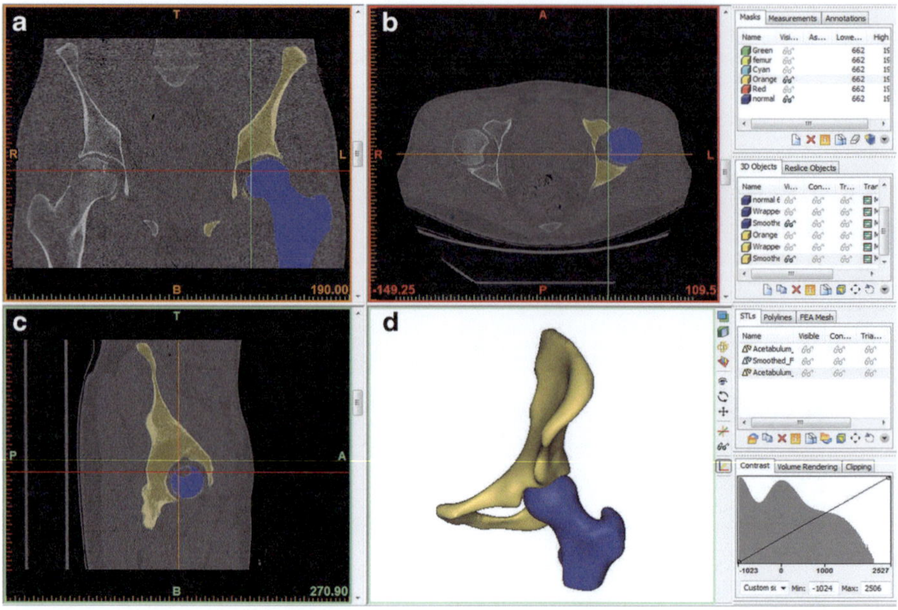

Fig. 6 Segmentation density masks of the femur and pelvis in (a) coronal plane, (b) axial plane, (c) sagittal plane. (d) 3-D rendering of both the density masks

2.2.1 Operations for the Quality Improvement of the Triangular Elements

The aim was to improve the quality of the triangular elements through different operations prior to autoremeshing of the 3-D models. The amount of detail of the femur was reduced by conservatively smoothening it without compromising the faithfulness of the model. A Laplacian (1st order) was used; a smoothing factor of 0.7 and 3 iterations were computed with compensation. However, the initial model contained a large number of triangular elements with different sizes. Thus, the triangles were resized and reduced to improve the quality of the model. A flip threshold value of 15, geometrical error of 0.2 and 5 iterations were chosen. Then, an autoremesh was carried out with a maximum edge length of the triangle limited to 5 mm. A quality threshold of 0.3, geometric error of 0.3 with control triangle edge operation on and an iteration of 4 was defined. However the mesh still consisted of group of small triangles. These were removed by using the quality preserving reduce triangles operation with a quality threshold of 0.3, maximum geometric error of 0.3, maximum edge length of 5 mm and with 3 iterations. The same operations were carried out on the pelvic girdle.

2.3 Surface Mesh to Volumetric Mesh

The head of the femur and the osseous bump at the head-neck junction was manually refined with the triangles smaller than 2.5 mm in order to get an accurate

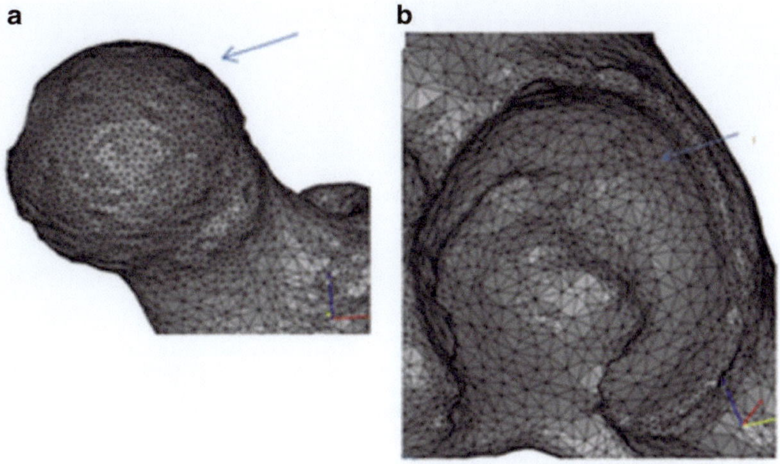

Fig. 7 (**a**) The head of the femur (*arrow*) was refined and acted as the slave surface. (**b**) The acetabulum (*arrow*) was coarsely meshed relative to the femur and was acted as the master surface

result at the contact interface (Fig. 7). The head of the femur was acting as the slave surface mesh whereas the acetabulum was behaving as the master surface mesh in order to respect the master and slave formulation in Abaqus. Moreover, the head of the femur was smoothened and refined relative to the pelvic girdle in order to prevent any sharp corners in the model that might cause convergence problems and to prevent any penetrations of master nodes into the slave surface from going undetected (Abaqus 6.9) (as explained in Sect. 2.6).

When the surface meshes of the femur and pelvic girdle satisfactorily passed the mesh quality test, they were saved as 'inp' files format and exported to Abaqus separately. There, the triangular elements were converted to tetrahedral elements. The femur and the pelvic girdle consisted of 125,376 and 169,483 tetrahedral elements, respectively.

2.4 Instancing and Location of the Hip Joint Centre

The femur was then instanced with the pelvic girdle using part/model instance option in Abaqus. In this study, a sphere was wrapped around and fitted on the femur to approximately locate the hip joint centre which was defined as a reference point (Fig. 8).

2.5 Material Property of the Finite Element (FE) Model

The bone was assumed to behave as a non linear elastic and an isotropic Poisson's ratio of 0.3 was used which is sufficient for the study of stress and strain

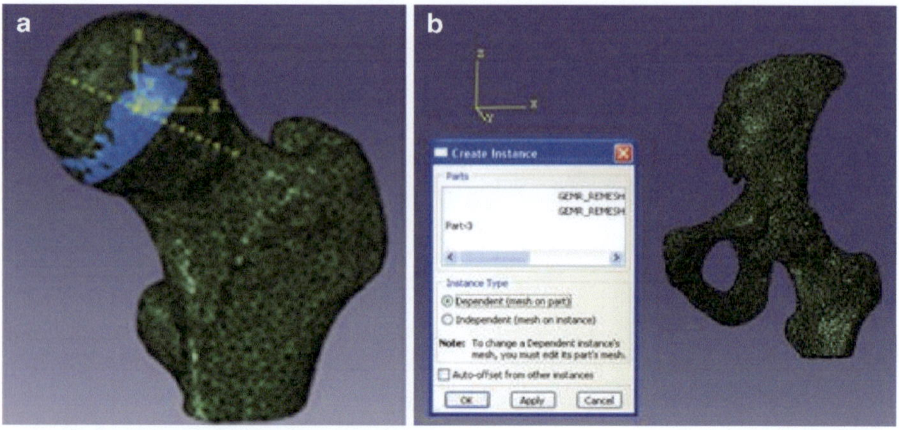

Fig. 8 (**a**) The hemispherical shell was wrapped, rotated and adjusted around the head of the femur till it fitted on the convex surface of the femur. (**b**) The femur and acetabulum was separately chosen and defined as instances. The autooffset from other instances was toggled on to assemble both the femur and acetabulum together

(Cowin 2001). An equivalent bone material property was defined for the finite element (FE) model with an average value of elastic modulus of 750 MPa and an average density of 1.281 g/cm^3.

2.6 *Contact Algorithm and Boundary Conditions*

In this study, the general automatic contact algorithm was defined in Abaqus. The contact between the acetabulum and the femur was assumed to be ideally frictionless with no bonding. As the acetabulum and femur comes into contact, the penetrations are detected and the contact constraints are applied according to the penalty constraint enforcement method (Abaqus 6.9).

A pure master-slave weighting was defined to resist only penetrations of slave nodes into master faces. Slave nodes are acted by contact forces proportional to the penetration distance in order to oppose the penetration. As a result, equal and opposite forces also act on the master surface and are distributed to its nodes (Abaqus 6.9).

The medial pelvic wall was pinned in the x, y and z directions (Fig. 9). The pelvic bone support of the acetabulum was non-rigid. This was important in order to make the pelvic bone compliant and thus facilitate attainment of continuous horseshoe shaped contact of a normal hip articulation (Russell et al. 2006). The centre of rotation of the femur was coupled with the whole section of the femur (Fig. 9).

During the translation and rotation operations, the centre of rotation was chosen and different physiological values were defined. This enabled the femur to move in

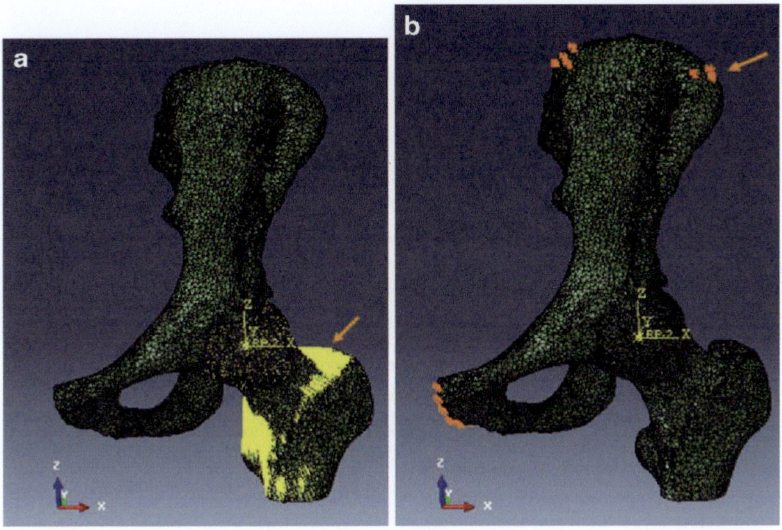

Fig. 9 (**a**) The centre of rotation was coupled (*arrow*) with the femur (**b**) The pelvic wall was pinned (*arrow*) at the corners with a non rigid bone support of the acetabulum

the predefined directions. However, due to lack of kinematic data from the patient prior to the arthroscopic surgery, the kinematics of the hip joint of the patient was estimated based on available literature. Physiological kinematics values that reported the range of motion of patients with symptomatic anterior hip impingement and normal hips were used as control (Clohisy et al. 2009; Kubiak-Langer et al. 2007) during definition of the values of rotation in Abaqus.

A provocative test such as anterior impingement test was simulated in order to investigate the range of motion when abutment of the osseous bump on the rim of the acetabulum occurred. In an anterior impingement test, the patient is placed supine on the examination table, and then the examiner passively flexes the hip to 90°, followed by forced adduction and internal rotation (Philippon et al. 2007). In this position, the anterior femoral neck approximates the anterosuperior acetabulum potentially causing the mechanical impingement that is thought to cause symptoms in FAI (Philippon et al. 2007).

A set of combinations of flexion, adduction and internal rotation was defined within the limit of physiological range of motion of a normal hip joint for three cases: healthy articular cartilage, and thinning of 2 mm and 4 mm of the articular cartilage. It has also been reported that the most critical motion for femoroacetabular impingement is internal rotation of the hip in 90° flexion (Ganz et al. 2008; Tannast et al. 2007b). Therefore, the femur was also flexed and internally rotated for the case of the healthy articular cartilage, 2 mm and 4 mm thinning of the articular cartilage. In fact, the anterior impingement test, and combined flexion and internal rotation were simulated till contact of the head-neck junction with the rim of the acetabulum was first recognised.

3 Results

3.1 *Kinematics of Impingement*

The results of the anterior impingement test are depicted in a bar chart for the three cases: healthy articular cartilage, 2 mm and 4 mm thinning of the articular cartilage (Fig. 10). Overall, the range of motion of the hip joint decreased with attenuation of the articular cartilage.

We observed that there was a decrease in flexion, adduction and internal rotation when the articular cartilage was attenuated by 2 mm. Similarly, there was a marked decrease in flexion and a slight fall in internal rotation, albeit with a rise in adduction when the articular cartilage was thinned by 4 mm. It was found that when the articular cartilage was becoming progressively worn out, the risk of impingement increased, as it was noted to occur earlier during flexion (Fig. 10).

The pathological hip joint was also flexed and internally rotated only (no adduction) for the above three cases. We found that bone-to-bone contact (impingement) occurred at lower values of flexion and internal rotation with thinning of the articular cartilage (Fig. 11).

3.2 *Location of Impingement*

A clock system or convention used in clinical practise was adopted to display the region of impingement on the acetabulum (Tannast et al. 2008). The acetabulum was divided into 12 sectors corresponding to a clock face. The clock was positioned in such a way that the 6 o'clock was located in the middle of the acetabular notch (Fig. 12).

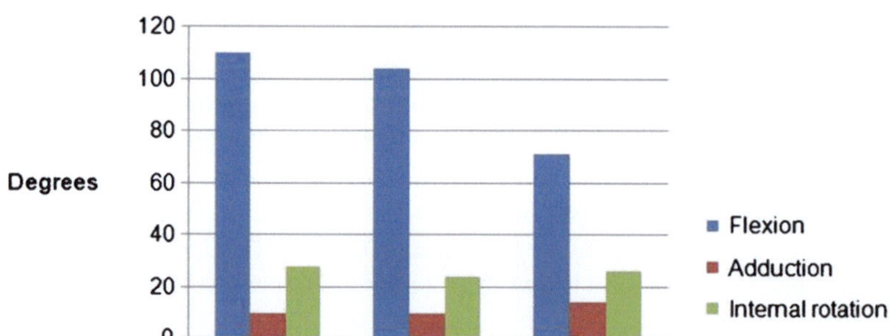

	Healthy	2 mm thinning	4 mm thinning
Flexion	110	104	71
Adduction	10	9.5	14
Internal rotation	28	23.8	26

Fig. 10 A bar chart shows the kinematics of impingement of the hip with healthy cartilage, 2 mm and 4 mm attenuated cartilage during the anterior impingement tests

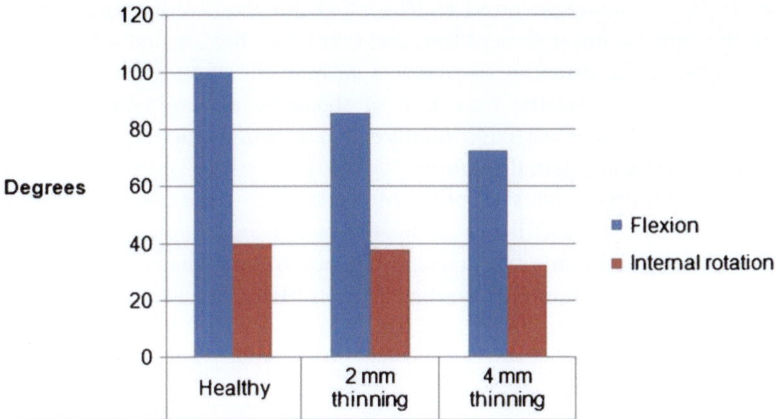

	Healthy	2 mm thinning	4 mm thinning
Flexion	100	85.5	72.7
Internal rotation	40	38	32.4

Fig. 11 A bar chart shows the kinematics of impingement of the hip with healthy cartilage, 2 mm and 4 mm attenuated cartilage during a combined flexion and internal rotation

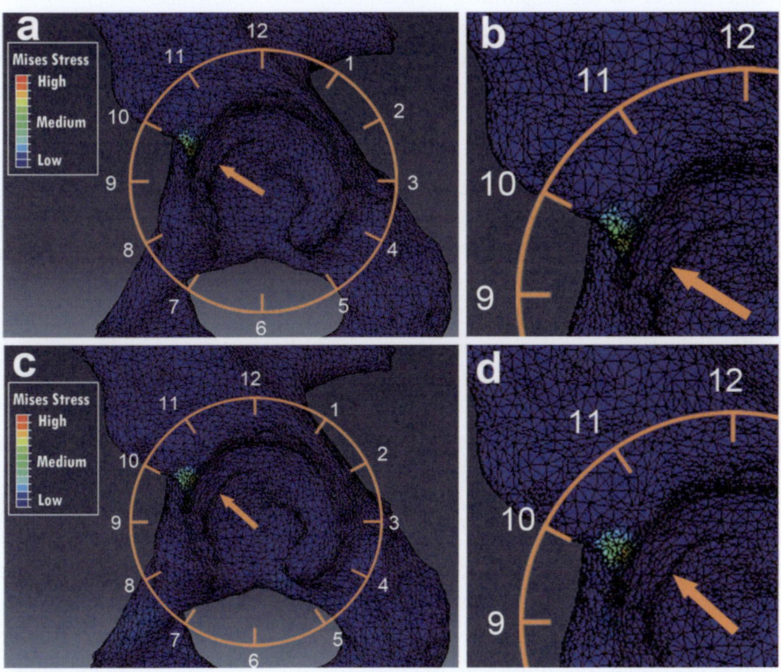

Fig. 12 (**a**) Impingement occurred at 10 o'clock for the healthy articular cartilage during combined flexion and internal rotation. The 6 o'clock was located in the middle of the acetabulum notch. (**b**) showing an enlarged view of the impingement zone for the healthy articular cartilage (**c**) Impingement occurred between 10 and 11 o'clock for the 2 mm attenuated articular cartilage during combined flexion and internal rotation. (**d**) showing an enlarged view of the impingement zone for the 2 mm attenuated articular cartilage

Bone-to-bone contact occurred at 10 o'clock for the healthy articular cartilage for both the anterior impingement test, and combined flexion and internal rotation. With thinning of the articular cartilage, a shift in the zone of impingement was observed. Fig. 12 shows the transition of the zone of impingement for 2 mm thinning of the articular cartilage relative to a healthy articular cartilage during combined flexion and internal rotation.

It was also observed that the spread of the bone-to-bone contact changed with thinning of the articular cartilage as indicated by the contour plot (Fig. 12). The red colour indicates maximum bone-to-bone contact or Von Mises stress whereas the blue colour represents minimum contact or Von Mises stress.

4 Discussion

Despite extensive clinical research on FAI, the effect of attenuated cartilage on the kinematics thereof is still largely unknown. The current study is aiming to track the motion of a cam-type femoral head with cartilage thinning of varying severity. The hypothesis defined at the beginning of the study was confirmed.

It was found that the kinematics of impingement decreased with thinning of the articular cartilage. The hip joint with the healthy cartilage impinged at 100° of flexion and 40° of internal rotation. On the other hand, the hip joint with 2 mm and 4 mm thinning of the articular cartilage impinged at 85.5° of flexion, 38.0° of internal rotation and 72.7° of flexion, 32.4° of internal rotation, respectively. This suggests that the risk of impingement is higher with thinning of the articular cartilage, as it occurs at lower flexion values. Thinning of the articular cartilage, as part of the osteoarthritic process, is usually accompanied by other anatomical aberrations, such as formation of osteophytes and/or capsular contractures. As those may cause further mechanical obstruction to motion, it may be postulated that, in reality, the arthritic hip with cam deformity is likely to impinge even earlier during the motion cycle.

Bone-to-bone contact occurred at 10 o'clock on the anterosuperior rim of the acetabulum for the healthy articular cartilage. However, there was a shift in the zone of impingement and maximum bone-to-bone contact with thinning of the articular cartilage. The maximum bone-to-bone contact occurred between 10 and 11 o'clock with 2 mm thinning of the articular cartilage during combined flexion and internal rotation. It was also observed that the spread of the impingement or bone contact on the rim of the acetabulum changed with thinning of the articular cartilage.

The impingement or abutment against the articular cartilage happened just before bone to bone contact occurred. Thus, the range of motion depicted for the kinematics of impingement may exceed the real values because the articular cartilage was not included in the finite element model. However, it was reported that motion is intra-operatively restricted by bony contact in the case of anterior femoroacetabular impingement (Ganz et al. 2003; Kubiak-Langer et al. 2007). Therefore, the results of the kinematics showed the first contact between the osseous bump and the rim of the acetabulum irrespective of the cleavage or delamination of the articular cartilage from the labrum that occurred just before.

The articular cartilage was not included in the finite element model at this stage because the interest of the study was to investigate the effect on the kinematics of the hip joint resulting from thinning or damage of the cartilage. It has been reported that the average thickness of the articular cartilage for the femur and the acetabulum in normal cadaveric hips was 2.83 mm and 2.97 mm, respectively (Wyler et al. 2007). In the presence of established advanced osteoarthritis, the resulting motion comprises both translation and rotation (Kubiak-Langer et al. 2007). This is why translation and rotation was defined in a single step in Abaqus to simulate the thinning of the 2 mm and 4 mm of the articular cartilage.

Indeed, when the axis of rotation was translated by 2 mm equally in all directions, this resulted in joint space narrowing. Hence, thinning of the articular cartilage by 2 mm and 4 mm is likely to represent moderate and advanced osteoarthritis, respectively. However, thinning was a bioengineering assumption and had no medical base. Therefore, the purpose of the current study was to investigate if there was a significant change in the kinematics of impingement when the axis of rotation shifts intra-operatively in an osteoarthritic hip.

Due to lack of kinematic data from the specific patient, the simulated kinematic values might deviate from the actual kinematics. There exist many possibilities of flexion, adduction and internal rotation or combined flexion and internal rotation that lead to bone-to-bone contact during the anterior impingement test (Tannast et al. 2008). In fact, the kinematic obtained in this study was one of the several possibilities that led to bone-to-bone contact or impingement. Whether this kinematic value represents the position of the hip when the patient was in pain is unknown. Nonetheless, the range of motion or combination defined was within the physiological range of motion of normal hips (Kubiak-Langer et al. 2007; Clohisy et al. 2009).

The model was pre-processed and meshed in Mimics prior to exporting to Abaqus. The anterosuperior osseous bump located on the head-neck junction was conservatively refined and smoothened without compromising the impairment of the faithfulness of the shape representation. Some authors have reported kinematics of impingement of the hip joint using computer-based approaches (Kubiak-Langer et al. 2007; Tannast et al. 2007a; Beaulé et al. 2005). In this study, a numerical analysis approach was adopted to investigate the kinematics of impingement.

Although there is enough concomitant gliding of the axis of rotation to move in space in terms of arthrokinematics, this is considered subtle and is frequently ignored (Levangie and Norkin 2001). Therefore, the centre of rotation was fixed during the simulation.

5 Conclusions

The kinematic data obtained is the first ever evidence demonstrating that the thickness of cartilage adversely affects the kinematics of a hip joint with cam-type FAI. This information may prove useful in understanding the changes occurring in such a hip, and may explain, at least in part, the rapid clinical deterioration of those patients, once arthritis develops.

Acknowledgments We thank the South West London Elective Orthopaedic Centre (SWLEOC) in the United Kingdom for having provided a Computed tomography (CT) scan and a magnetic resonance (MR) arthrogram of a patient for the purposes of the investigation.

References

Abaqus User's Manual, Version 6.9 (2011) Dassault Systemes Simulia, Providence, Rhode Island, USA. www.simulia.com

Beaulé PE, Zaragoza E, Copelan N, Dorey F (2005) Three-dimensional computed tomography of the hip in the assessment of femoroacetabular impingement. J Orthop Res 23:1286–1292

Clohisy JC, Knaus ER, Hunt DM, Lesher JM, Harris-Hayes M, Prather H (2009) Clinical presentation of patients with symptomatic anterior hip impingement. Clin Orthop Relat Res 467:638–644

Cowin SC (2001) Bone mechanics handbook, 2nd edn. CRC Press LLC, New York

Dagenais S, Garbedian S, Wai EK (2009) Systematic review of the prevalence of radiographic primary hip osteoarthritis. Clin Orthop Relat Res 467(3):623–637

Ganz R, Parvizi J, Beck M, Leunig M, Notzli H, Siebenrock KA (2003) Femoroacetabular impingement: a cause for osteoarthritis of the hip. Clin Orthop Relat Res 417:112–120

Ganz R, Leunig M, Leunig-Ganz K, Harris WH (2008) The etiology of osteoarthritis of the hip: an integrated mechanical concept. Clin Orthop Relat Res 466:264–272

Ingvarsson T (2000) Prevalence and inheritance of hip osteoarthritis in Iceland. Acta Orthop Scand Suppl 298:1–46

Kubiak-Langer M, Tannast M, Murphy SB, Siebenrock KA, Langlotz F (2007) Range of motion in anterior femoroacetabular impingement. Clin Orthop Relat Res 458:117–124

Levangie PK, Norkin CC (2001) Joint structure and function: a comprehensive analysis, 3rd edn. F.A Davis Company, Philadelphia

Mimics User Manual, version 13.1 (2011) Materialise, Leuven, Belgium. www.materialise.com

Philippon MJ, Maxwell RB, Johnston TL, Schenker M, Briggs KK (2007) Clinical presentation of femoroacetabular impingement. Knee Surg Sports Traumatol Arthrosc 15:1041–1047

Russell ME, Shivanna KH, Grosland NM, Pedersen DR (2006) Cartilage contact pressure elevations in dysplastic hips: a chronic overload model. J Orthop Surg Res 3:1–6

Tannast M, Kubiak-Langer M, Langlotz F, Puls M, Murphy SB, Siebenrock KA (2007a) Non-invasive three-dimensional assessment of femoroacetabular impingement. J Orthop Res 25:122–131

Tannast M, Siebenrock KA, Anderson SE (2007b) Femoroacetabular impingement: radiographic diagnosis-what the radiologist should know. AJR Am J Roentgenol 188:1540–1552

Tannast M, Goricki D, Beck M, Murphy SB, Siebenrock KA (2008) Hip damage occurs at the zone of femoroacetabular impingement. Clin Orthop Relat Res 466:273–280

Wyler A, Bousson V, Bergot C, Polivka M, Leveque E, Vicaut E, Laredo J-D (2007) Hyaline cartilage thickness in radiographically normal cadaveric hips: comparison of spiral CT arthrographic and macroscopic measurements. Radiology 242:441–449

Image Based Model Development and Analysis of the Human Knee Joint

Dennis John, Dinesh Pinisetty, and Nikhil Gupta

Abstract Developments in medical imaging and finite element analysis techniques have made it possible to conduct personalized studies on patients. The field of medical implants is especially benefitting from these advancements, where patient specific geometries can be created and analyzed. The present work is focused on using image based techniques for construction of solid models of human knee joints for finite element analysis. Accurate 3D solid models of the human cadaveric knee joint are developed based on a sequence of high resolution MRI images obtained from a Siemens 7T machine. The approach involves identification of various components of the knee joint such as the femur, tibia, femoral and tibial cartilage, and menisci of the tibio-femoral knee joint; construction of a 3D model; smoothing the geometries; meshing of geometry; and then performing finite element analysis. The focus of the present work is on understanding the effect of menisci on the stress and strain distribution in the knee joint. Availability of such image based modeling and analysis methods would help in designing effective meniscal implants.

Keywords Meniscus • Knee joint • Magnetic resonance imaging

D. John
Composite Materials and Mechanics Laboratory, Mechanical and Aerospace Engineering Department, Polytechnic Institute of New York University, Six MetroTech Center, Brooklyn, NY 11201, USA

General Dynamics Electric Boat, 75 Eastern Point Road, Groton, CT 06340, USA

D. Pinisetty • N. Gupta (✉)
Composite Materials and Mechanics Laboratory, Mechanical and Aerospace Engineering Department, Polytechnic Institute of New York University, Six MetroTech Center, Brooklyn, NY 11201, USA
e-mail: ngupta@poly.edu

D. Iacoviello and U. Andreaus (eds.), *Biomedical Imaging and Computational Modeling in Biomechanics*, Lecture Notes in Computational Vision and Biomechanics 4,
DOI 10.1007/978-94-007-4270-3_4,

1 Introduction

Analysis of complex systems such as human knee joints has tremendously benefitted from the recent developments in three dimensional (3D) imaging and finite element analysis (FEA) techniques. Presence of different materials, complex shapes, and variation in mechanical properties in human knee joints makes the analysis very challenging. Success of FEA strongly depends on two factors: (a) availability of solid models that accurately define the geometry of the knee joint and (b) availability of mechanical properties of each material that appears in the structure. There is remarkable progress in both these directions in the recent years. While the initial FEA efforts relied upon simplified geometries, such as assuming bones to be of tubular shape, a combination of 3D imaging techniques and advanced data processing methods have resulted in development of solid models that capture the structural details of actual joints. Similarly, the field of mechanical property measurement has significantly advanced in recent times. Availability of nanomechanical test systems has allowed localized measurements of mechanical properties of bones, which are more precise than the macroscale test data (Rho et al. 2002).

This work is focused on image based model development for FEA. There are several methods to create a 3D model that is useful for conducting FEA (Haut Donahue et al. 2002; Blankevoort et al. 1991; Yang et al. 2010; DeFrate et al. 2004; Hao et al. 2007; Bendjaballah et al. 1997; Vadher et al. 2006; Trilha et al. 2009; Périé and Hobatho 1998; Baldwin et al. 2009; Guo et al. 2009). Among the available medical imaging techniques computed tomography (CT) scans, magnetic resonance images (MRI), ultrasound, and laser digitizers are capable of providing outputs that can be converted to solid models. Creating models from CT or MRI is often time-consuming but can provide very accurate models with fine structural details (Cohen et al. 1999). In general, the benefit of MRI over CT scan is the ability to see the soft tissues of the joint which defines their application domains (Madelin et al. 2010).

Early studies used axisymmetric models, which compromised the solution accuracy (Vadher et al. 2006). This is because axisymmetric models poorly represent the three dimensional geometry of the femur-menisci-tibia complex (Haut Donahue et al. 2002). Apart from the femur and tibia bones, accurate geometries of the cartilage layers and menisci are important for the studies related to knee implants and arthritis (Bendjaballah et al. 1995).

The knee undergoes multiple physiological conditions such as sliding, rolling, and traction (Walker et al. 1997). The meniscus is important in load transmission, shock absorption, proprioception, and improvement of stability and lubrication (Vedi et al. 1999). The menisci distribute contact forces over the articular surfaces by increasing the contact surface area in the joint (Peña et al. 2005). This avoids stress concentrations on the surface of articular cartilage. Studying the effects of loading on the knee can help in taking preventative measures to avoid damage. Modeling the menisci correctly is important for the creation of an appropriate FE model of the knee joint.

In United States an estimated 50 million people were diagnosed with arthritis between 2007 and 2009 (Cheng et al. 2010a). A projected population of 67 million people is likely to be diagnosed with arthritis by 2030 (Hootman and Helmick 2006). Osteoarthritis (OA), which is a degenerative disease of the articular cartilage (Sharma et al. 2001), categorized as the structural and functional failure of synovial joints. OA persists when the dynamic equilibrium between the breakdown and repair of joint tissues is overwhelmed (Garstang and Stitik 2006). It has been shown that local biomechanical factors may affect initiation and progression of OA (Yang et al. 2010). It can lead to limited mobility, pain, and joint deformity (Arokoski et al. 2000). Local altered joint biomechanics that can lead to OA include ligamentous laxity, malalignment, impaired proprioception, and muscle weakness (Loeser and Shakoor 2003). Other studies have also shown that damage to the meniscus or partial meniscectomy can lead to OA (Zielinska and Haut Donahue 2006), which could be due to a change in the stress distribution patterns on articular cartilage (Shirazi and Shirazi-Adl 2009). MRI can be helpful in early detection by assessing cartilage health (Regatte and Schweitzer 2008).

Finite element analysis is helpful to study such problems and produces accurate results in terms of stress, strains, and stress concentrations (Vadher et al. 2006). Haut Donahue et al. concluded that a solid model is a viable option for modeling soft tissues to study the contact behavior in the knee joint (Haut Donahue et al. 2002). The assumption that cartilage behavior is linearly elastic instead of biphasic was found to be appropriate for short loading times (Haut Donahue et al. 2002). Previous studies have shown that increased body mass index can increase the risk of OA progression because of higher mechanical loading at the knee joint (Yang et al. 2007). A number of studies have analyzed the effect of meniscal tears and meniscectomies by comparing their results with those of healthy meniscus (Peña et al. 2005; Zielinska and Haut Donahue 2006; Shirazi and Shirazi-Adl 2009; Vaziri et al. 2008). In a healthy knee, uncovered tibial cartilage is loaded at the central regions near the tibial eminence. The areas that are covered by meniscus experience lower compressive stress (Bendjaballah et al. 1995).

The effects of meniscectomy have been extensively studied previously (Song et al. 2008; Peña et al. 2006, 2008). Results have shown that partial or total meniscectomy often leads to osteoarthritic degenerative changes in the articular cartilage of the femur and tibia (Vadher et al. 2006). Zielinska and Haut Donahue (2006) found that removing parts of the medial meniscus did not usually alter the location of maximum contact pressure, but increased the magnitude of contact pressure. Specifically, Shirazi and Shirazi-Adl (2009) found that partial meniscectomy shifts the transfer of applied compression away from the resected meniscus onto the nearby cartilage. When the meniscus is removed stress is relieved from the areas that were previously covered by the meniscus. In addition, the areas that were uncovered by meniscus become over-stressed, reducing the contact area, and concentrate the contact stresses on the cartilage (Bendjaballah et al. 1995). Meniscal tears are just as important as cartilage damage to look for in MRIs (Alatakis and Naidoo 2009). Meniscal tears can lead to OA, but OA can also lead to a meniscal tear (Englund et al. 2009; Englund 2008). From these studies it appears that including menisci in the model is necessary for increasing the accuracy of the analysis. The goal of this study is to create a 3D model of the tibio-femoral joint, which encompass bones, cartilage, and menisci. The eventual use of this model will be to analyze the effects of damaged menisci and those of meniscal implants.

2 Materials and Methods

The specimen for the modeling was acquired from the left knee of a cadaveric specimen of a normal male of 60 years of age. The dissection of the specimen is depicted in Fig. 1. The MRI of the specimen was taken before dissection. The specimen was placed in a test rig to force it to be in 15 degrees of flexion. A custom built apparatus was setup to make sure that the joint was under compression to get accurate images (Walker et al. 2009, 2011). Such a setup ensures that the femur applies compressive load onto the joint.

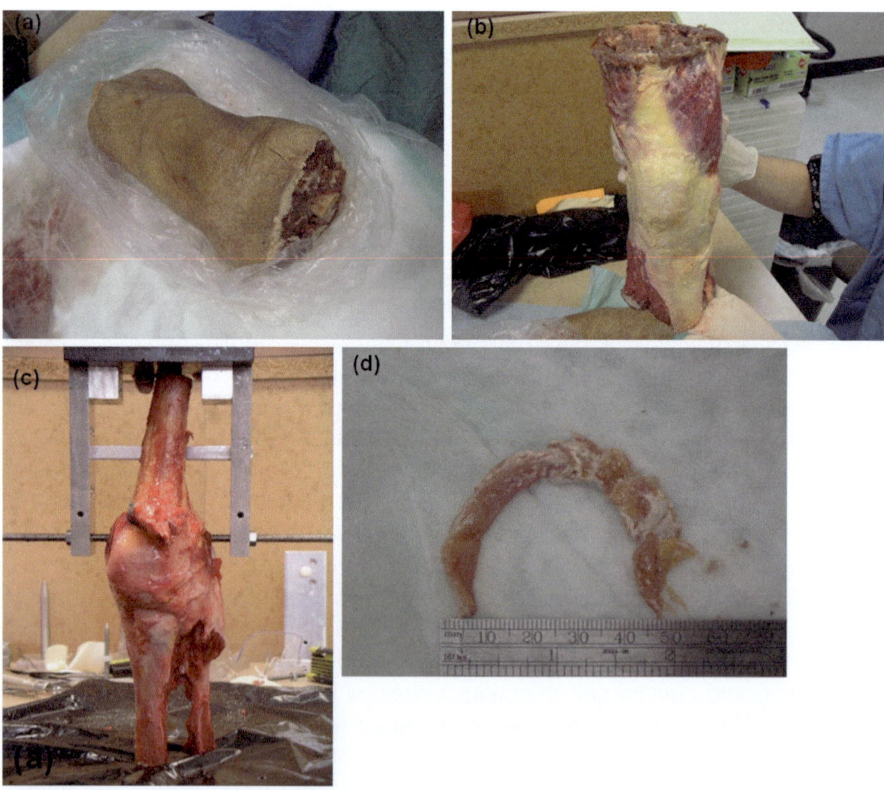

Fig. 1 (**a**) A cadaveric knee specimen used for MRI, (**b**) specimen after removal of skin, (**c**) specimen after further dissection, and (**d**) the medial meniscus

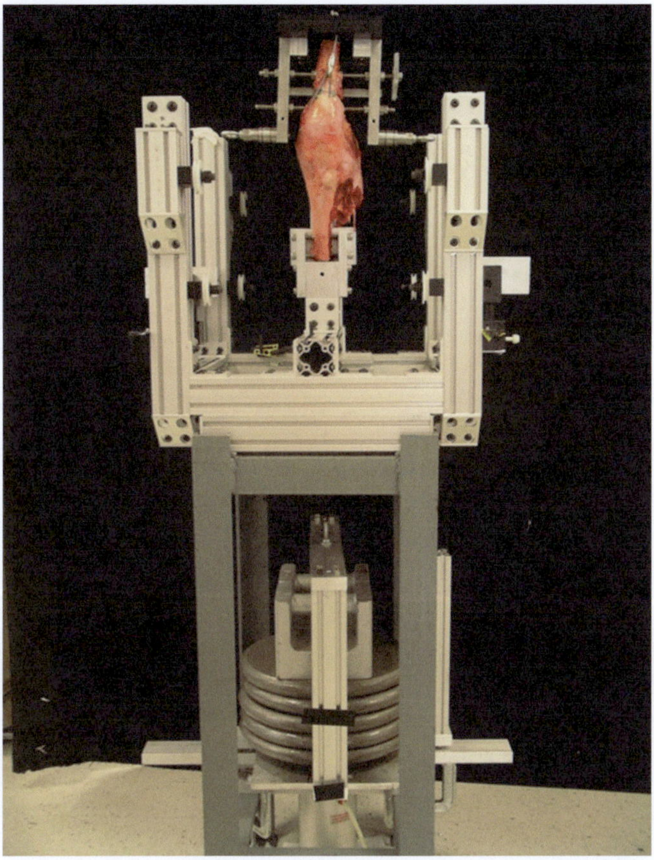

Fig. 2 Mechanical testing performed on cadaveric human knee specimen under compression

Experimental testing was performed on various cadaveric knee specimens (Chaudhary et al. 2011). The purpose of the testing was to investigate the role of the anterior cruciate ligament (ACL), posterior cruciate ligament (PCL) and medial meniscus in knee stability. During the test, tibia was fixed vertically and the femur was allowed to move in all six degrees of freedom as shown in Fig. 2. This allowed the femur to locate in a physical position and the displacement was relative to that of the tibia.

The findings of experimental testing are described in Chaudhary et al. (2011). Some general trends which were observed are: (1) once the ACL was resected there was posterior displacement of the femur relative to the tibia under all the test conditions, (2) once the medial meniscus was removed there was an increase in posterior displacement, (3) the removal of the medial meniscus increased joint laxity and decreased joint stability. There were even some instances where the knee specimens tested were completely dislocated after the removal of the medial meniscus under certain shears and torques. From the mechanical testing

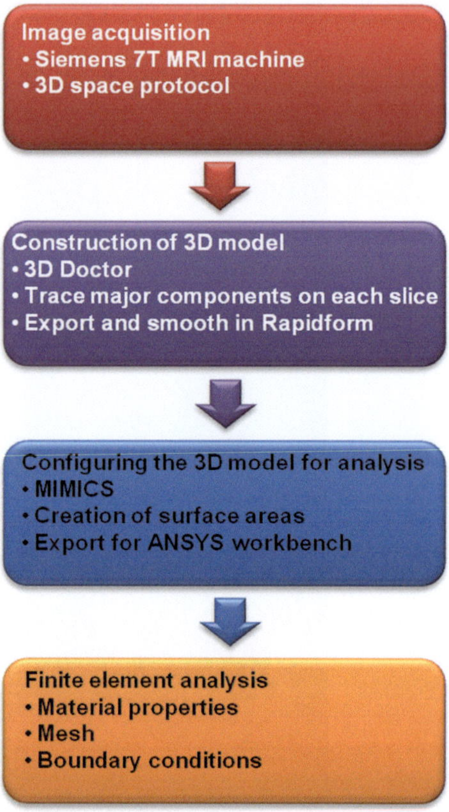

Fig. 3 Steps involved in creation of a solid model and conducting finite element analysis

performed it can be concluded that the ACL is important in keeping the femur from displacing posteriorly under the loading conditions on the knee (Chaudhary et al. 2011). The medial meniscus was a secondary stabilizer which experiences higher stresses under which damage can occur and cartilage wear can increase (Robertson et al. 2003).

The work flow in the creation of the model for FEA is summarized in Fig. 3. Each step in the process flow will be described in the following sections.

2.1 MRI Image Acquisition

Magnetic resonance imaging (MRI) is a widely used imaging technique due to its non-invasive evaluation of soft tissues. A basic comparison of various commonly used medical imaging techniques is provided in Table 1.

The MRI image acquisition was performed at NYU Langone Medical Center using Siemens 7T scanner (Siemens, Malvern, PA) shown in Fig. 4.

Table 1 Comparison of different radiography techniques

Category	X-ray	CT-scan	MRI	Ultrasound
Source	Radiation	Radiation	Electromagnetic waves	Ultrasonic waves
Image	2D	3D	3D	3D
Soft tissues	No	No	Yes	Yes
Image quality	Low	High	Very high	High
Cost	Low	High	Very high	High

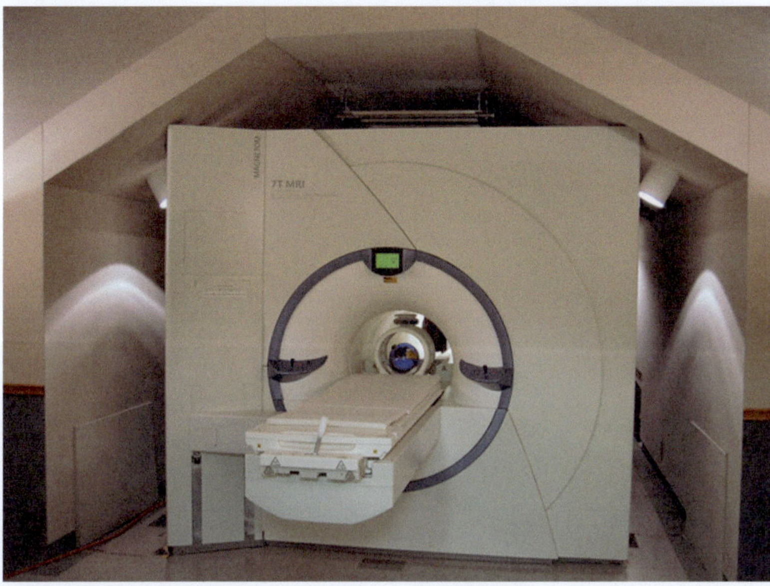

Fig. 4 Photograph of Siemens 7T scanner (Courtesy Dr. R.R. Regatte)

Although the Siemens 7T scanner is not yet widely used due to its associated costs and lack of FDA approval for scanning live patients, the increased static field strength improves signal-to-noise ratio and enhances image quality (Pakin et al. 2006; Chang et al. 2010). Figure 5 depicts the knee joint MRI images obtained from Siemens 3T and Siemens 7T scanners. The image obtained from the 7T scanner has a better resolution. High resolution MRI images allow capturing finer geometrical details and smaller objects and keeping track of the musculoskeletal soft tissues which are hard to be identified by commonly used radiography techniques (Regatte and Schweitzer 2007).

In the present study a 3D-SPACE (Sampling Perfection with Application optimized Contrasts using different flip angle Evolutions) protocol in 0.5 mm increments was employed for MRI imaging. The benefit of this protocol is that it allows 3D isotropic resolution, which enables visualizing the menisci, ligaments, and soft tissue. The acquisition parameters in the 3D-SPACE were: flip angle 160°, echo time 27 ms, repetition time 2,010 ms, bandwidth 86.25 Hz/pixel, matrix size 320×320 pixels, and voxel size $0.47 \times 0.47 \times 0.5\ \text{mm}^3$. After a steady state was reached the total acquisition time was 8:20 min. The number of slices for slab was 192.

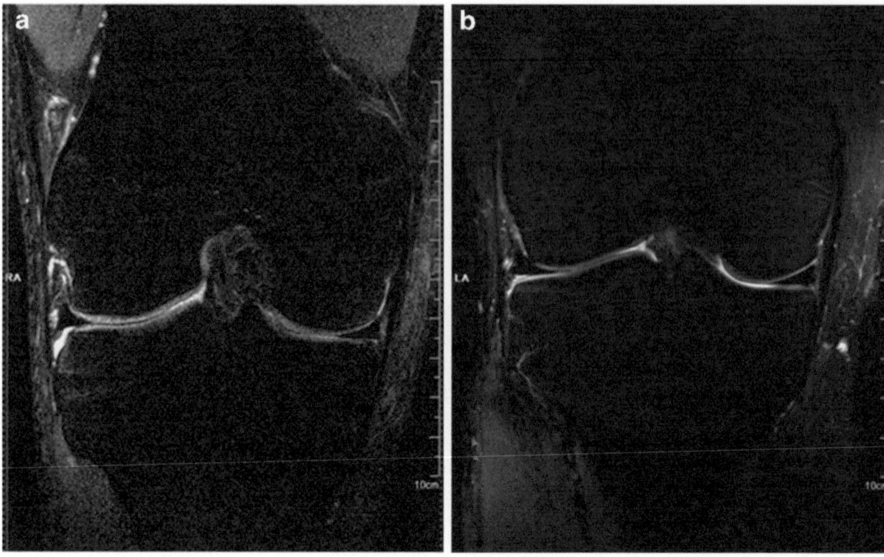

Fig. 5 Comparison of the knee joint MRI image obtained from (**a**) 3T and (**b**) 7T scanners, manufactured by Siemens

Fig. 6 MRI slice extracted from knee joint with acquisition parameters as flip angle 160°, echo time 27 ms, repetition time 2,010 ms, bandwidth 86.25 Hz/pixel, matrix size 320 × 320 pixels, voxel size $0.47 \times 0.47 \times 0.5$ mm^3

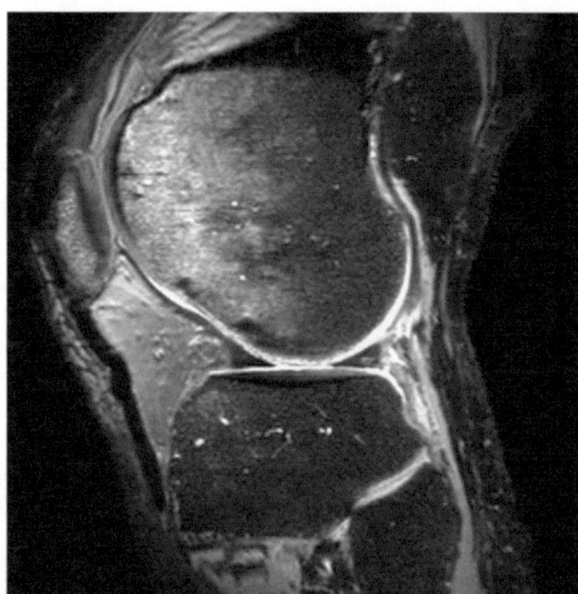

This gives a total specimen width of 96 mm. Starting with the MRI slice image and by tracking the boundaries, all the knee components (femur, tibia, femoral cartilage, tibial cartilage and menisci) are identified. The MRI slice extracted by employing the aforementioned acquisition parameters is depicted in Fig. 6. To identify the menisci

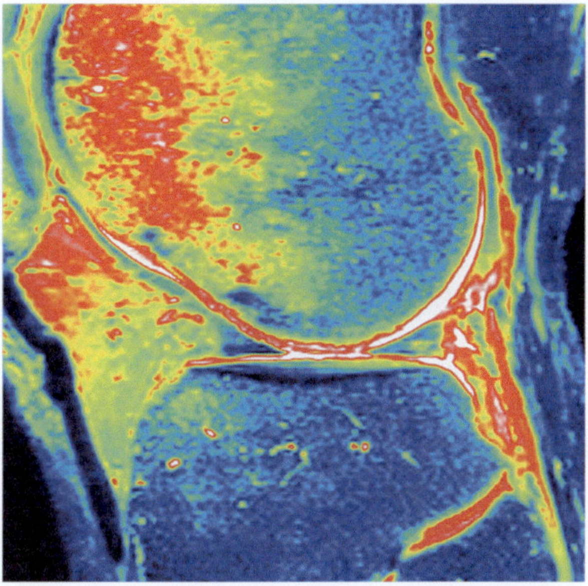

Fig. 7 Pseudo-color image palette applied to the MRI to clearly identify the meniscus. The meniscus can be visualized as the *blue region* between the femur-tibia joint

clearly, pseudo color images can also be extracted from MRI. Figure 7 depicts a pseudo color image extracted from MRI, where blue region between femur and tibia can be identified as menisci. The MRI image slices are then imported into solid modeling software to reconstruct the geometry of the tibio-femoral joint contact area.

2.2 *Construction of 3D Model*

The DICOM output files of the MRI image slices were imported into 3D Doctor (Able Software Corp., Lexington, MA) for geometrical construction of the components of the knee joint. 3D Doctor is a tool for advanced 3D modeling and imaging processing for MRI, CT, PET, microscopy, and several other scientific and industrial imaging applications (Ni et al. 2008; Tang et al. 2009; Fan et al. 2008; Ferraz et al. 2011; Busse et al. 2008). It creates a solid body by virtually connecting all the slices that have been traced, similar to the "loft" function in CAD software. The femur, tibia, femoral cartilage, tibial cartilage, and menisci were constructed using slices in the sagittal plane. Each slice must be segmented manually to create a continuous surface (Baldwin et al. 2010). The femur and tibia are the simplest to identify. The dark gray regions signify bone in the MRI output. As an example, it can be seen in Fig. 8 that the slices are segmented manually and the boundary of the femur is traced by a blue line and that of tibia is traced by a green line. The cartilage can be identified by the white regions around the gray regions representing bones. The menisci can be identified as the dark region at the center of the joint. Tracings were performed for femoral cartilage, tibial cartilage

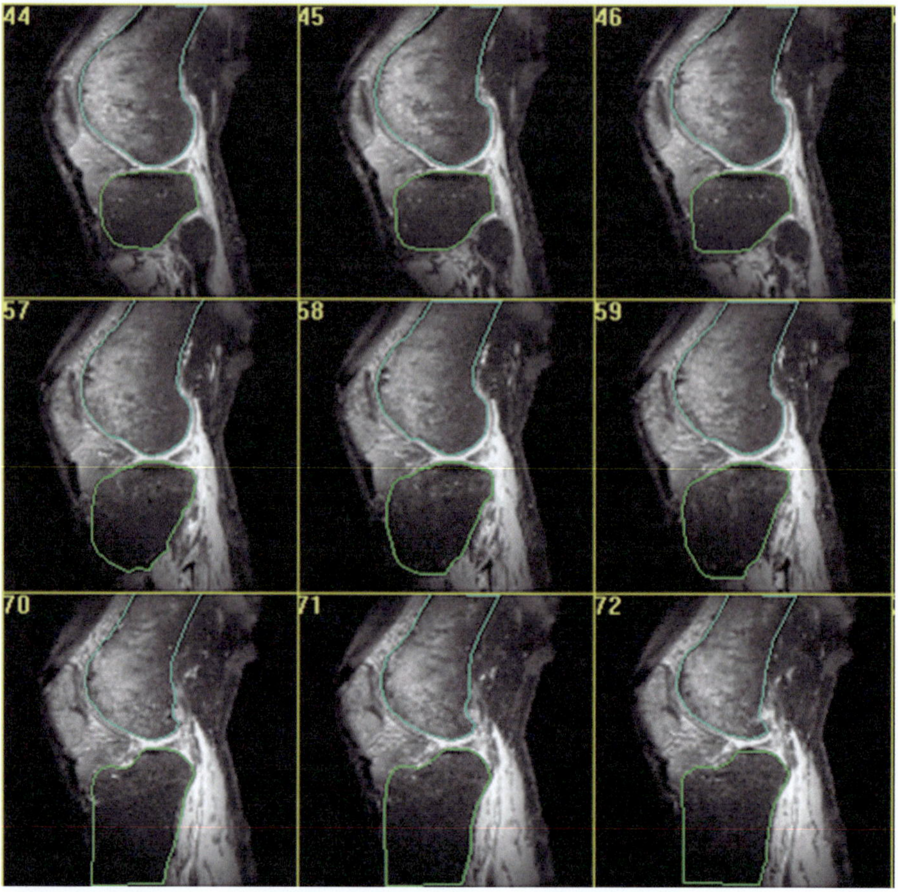

Fig. 8 Boundary tracing of femur and tibia bones within a knee joint

and menisci (see Fig. 9). The geometries were constructed by controlled tracing of the boundary of each component on each slice and then a solid body is created. Figure 10 depicts the solid model created in 3D doctor. Once the initial 3D model is created it must be inspected for geometrical anomalies such as holes and any remaining imaging artifacts.

2.3 *Processing of Solid Model*

Finite element analysis can be efficiently performed on the models after the imaging artifacts are removed while preserving the geometric details. The model in 3D Doctor was exported to Rapidform XOR (INUS Technology Inc., Seoul, South Korea) and all five components (femur, tibia, femoral and tibial cartilages,

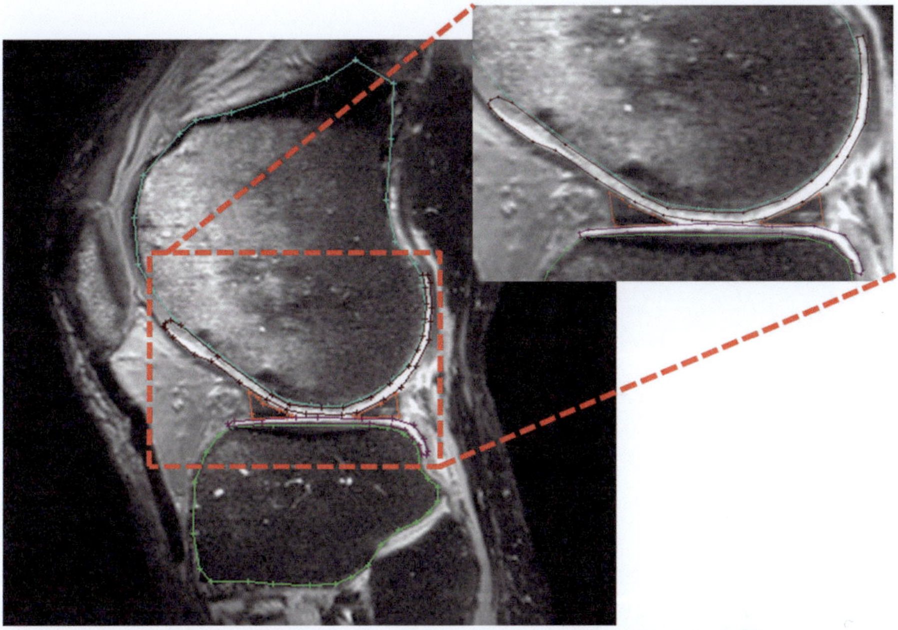

Fig. 9 Boundary tracing of femur, tibia, cartilages, and menisci. Femur and tibia are traced by *blue* and *green lines*, respectively. Cartilages and menisci are traced by *brown* and *orange lines*, respectively

and meniscus) were smoothed. Rapidform checks for any irregularities present in the solid model created in 3D Doctor and smoothens them (Shin et al. 2007; Cheng et al. 2010b). The smoothing options that were selected in the present study are maximum strength and maximum smoothness level with shrinkage compensated to avoid large geometrical changes. It is important not to smooth geometries more than necessary to avoid a compromise in the solution accuracy. Figure 11 depicts the Rapidform smoothed 3D solid model imported from 3D Doctor. Figure 12 explicitly shows the different views of the knee joint. The posterior and side views of the medial side of the joint are shown in Fig. 12a and b, respectively. Figure 12c depicts the proximal tibia with cartilage and meniscus. These one time maximum smoothed components were again exported as STL files from Rapidform into MIMICS.

The Rapidform smoothed solid model is imported into MIMICS (Materialise, Leuven, Belgium) to perform further smoothing and preparation of the model for FEA using ANSYS (Liu 2011). All the knee components (femur, tibia, cartilages, and menisci) were smoothed for 100 iterations with a smoothing factor of 0.4. A higher level of smoothing can be more effective but the geometry may be significantly changed. The level of 0.4 was selected because if the menisci and cartilage were smoothed any further, ANSYS would not register a thickness for these components.

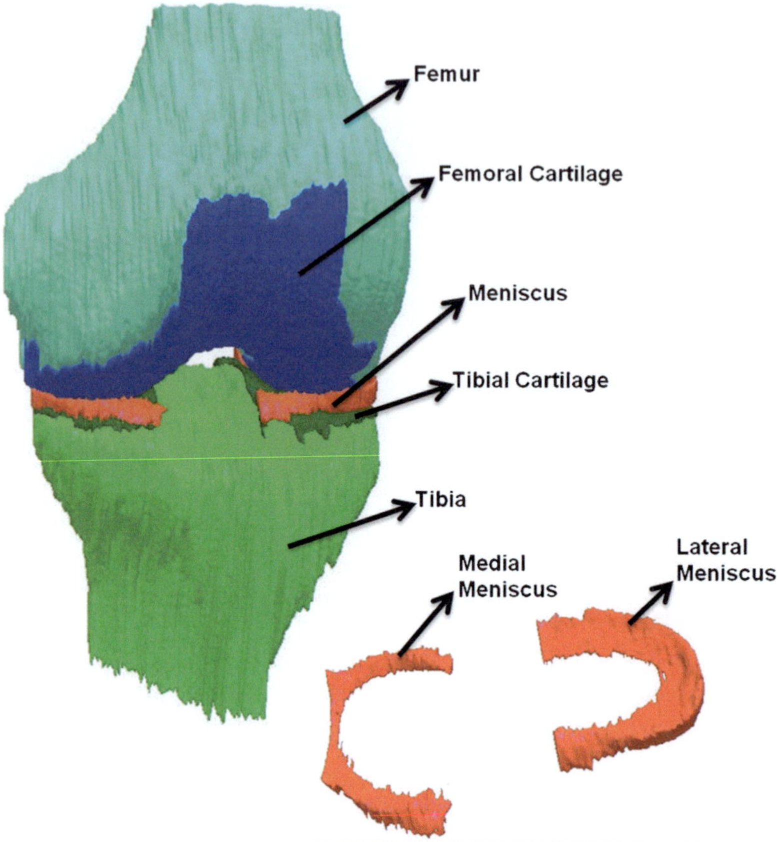

Fig. 10 Solid model of knee joint prepared using 3D Doctor is shown with all the components (femur, tibia, cartilages, and menisci). A *zoomed in view* of menisci is shown as an *inset* in the figure

There is a significant effect of the number of iterations on smoothing of the geometry as shown in Fig. 13. It can be observed from Fig. 13 that smoothness of the geometry increases with the number of iterations. Such a smoothed geometry decreases the number of nodes and elements required in the discretized finite element model and reduces computational time. Figure 14 depicts the anterior, posterior and side views of the solid model of the knee joint with all its components after smoothing using MIMICS.

Once smoothing was performed, MIMICS Remesh was used to prepare the solid bodies for ANSYS. MIMICS transforms the solid bodies into triangular areas for simpler transfer to FE modelers (Gíslason et al. 2010; Magne 2007). When these bodies of areas are created, the number of area elements can sometimes be excessive. To simplify this problem there are various functions within MIMICS Remesh. Two commonly used functions are "quality preserving reduce triangles"

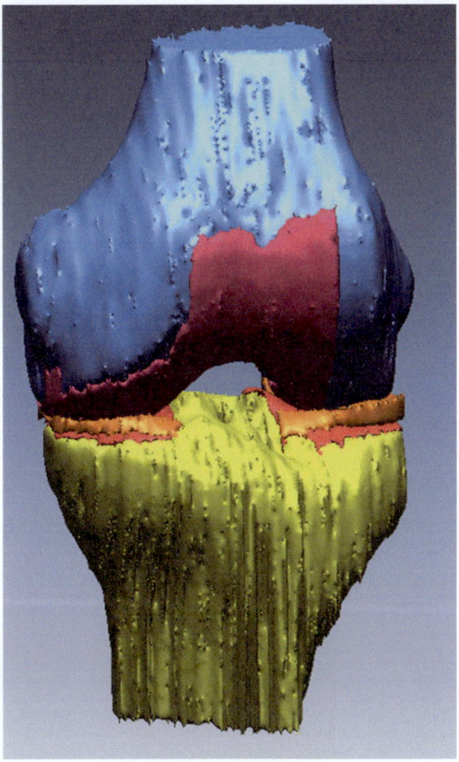

Fig. 11 Solid model of knee joint after smoothing once in Rapidform

and "reduce triangles". When the reduce triangles function is used, it significantly cuts down the number of areas but the quality of the area elements could be compromised. This function should not be used loosely. Although the quality preserving reduce triangles function retains high quality area elements, it does not reduce the number of triangles by as much of a factor as the reduce triangles function. The final mesh qualities in the present study for the various components were as follows- Femur: 100% high quality; tibia: 99% high quality, 1% medium quality; femoral cartilage: 96% high quality, 4% medium quality; tibial cartilage: 94% high quality, 6% medium quality; menisci: 97% high quality, 3% medium quality. These components were then exported as ANSYS area files (with an extension of *.inp*).

The exported files are loaded in ANSYS Mechanical APDL and by using the function "read input from" these files are converted into IGES files. The option "offset element numbers" must be used to avoid overlapping of element and node numbers. ANSYS Workbench was used for meshing and analysis since it is more user friendly. Figure 15 depicts the anterior view and posterior view of the final knee joint model.

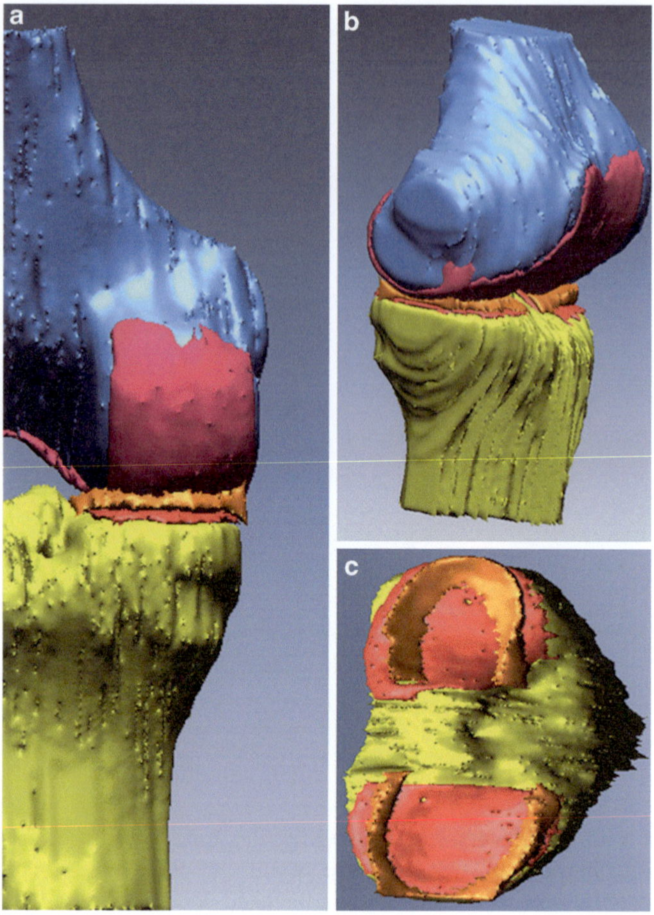

Fig. 12 (**a**) *Posterior view* and (**b**) *side view* of the medial side of the knee joint, and (**c**) *top view* of the meniscus, tibial cartilage, and tibia

3 Results and Discussion

The knee joint model, with all its components exported into ANSYS workbench, was meshed for FEA. The model was meshed producing components with the following number of elements and nodes: femur: 24,895 elements and 42,982 nodes, tibia: 25,917 elements and 44,398 nodes, femoral cartilage: 31,718 elements and 59,300 nodes, tibial cartilage: 15,725 elements and 31,125 nodes and menisci: 13,580 elements and 25,448 nodes. A coarse sized mesh was selected. The bones were considered to be elastic solids. The femoral and tibial cartilages were considered to be linear elastic and isotropic. However, if better mechanical property data is available, then it can be used in the analysis. The menisci were considered to be transversely isotropic elastic solids. Detailed mechanical properties are listed in Table 2.

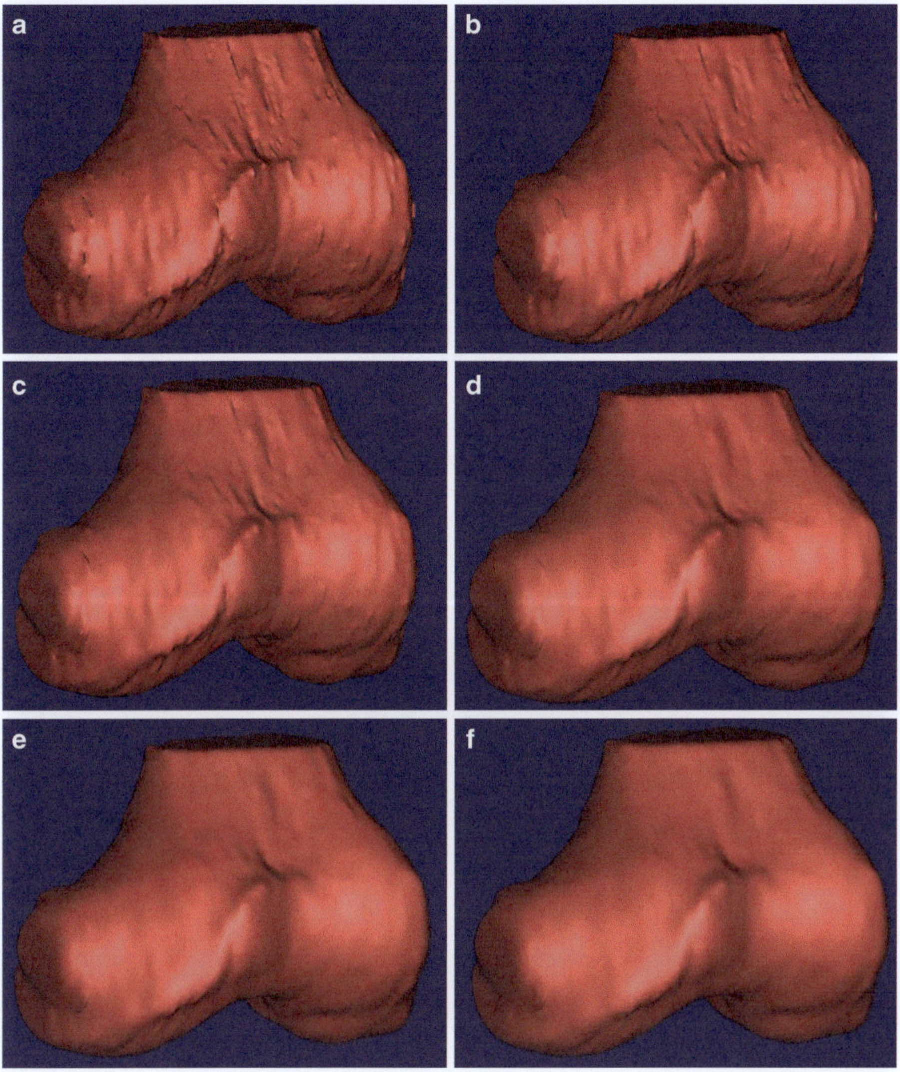

Fig. 13 Effect of the number of iterations on smoothing of the femur in MIMICS, (**a**) no smoothing and after (**b**) 10, (**c**) 25, (**d**) 50, (**e**) 75, and (**f**) 100 smoothing iterations

Femoral cartilage and tibial cartilage were glued to the surface of the femur and tibia, respectively. In total, four contact pairs were set up: femoral cartilage and tibial cartilage, femoral cartilage and meniscus, tibial cartilage and tibia, and tibial cartilage and meniscus. These contact pairs were defined to be frictionless. Figure 16 depicts the four contact pairs after their meshing in ANSYS Workbench. Meniscal horn attachments were modeled as linear springs with longitudinal stiffness of 2,000 N/mm (Haut Donahue et al. 2003). The meniscal horns were attached

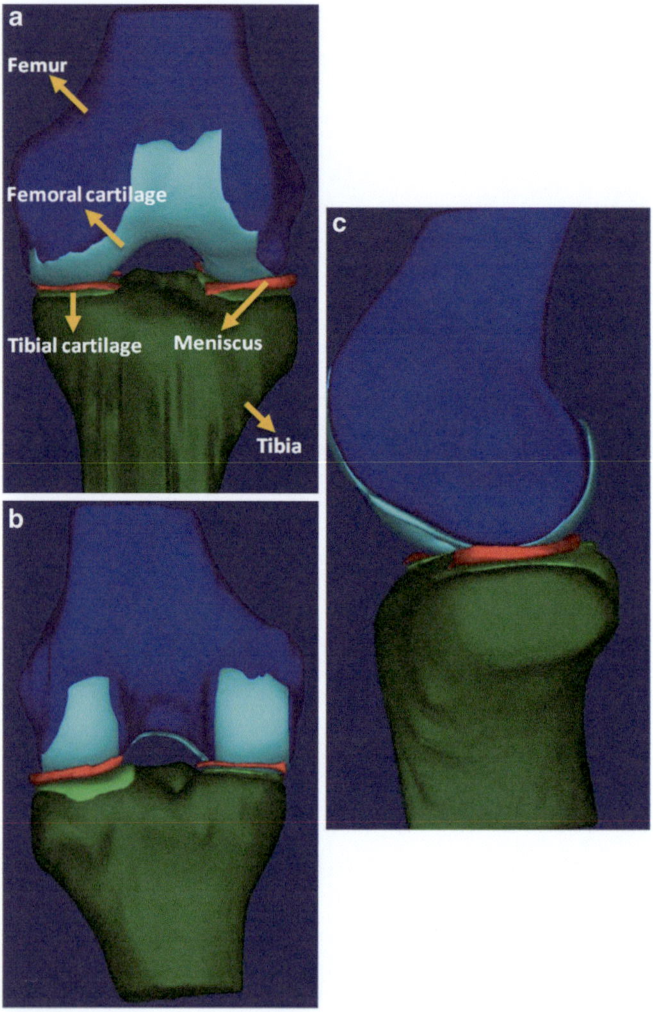

Fig. 14 Solid model of the final knee joint: (**a**) *anterior*, (**b**) *posterior*, and (**c**) *side view*

from both ends of the lateral and medial menisci to the surface of the tibial cartilage. The element types for the model include: SOLID187 and SURF154 to model the elastic components and COMBIN14 to model the linear springs for the meniscal attachments. Contact was modeled using elements CONTA174 and TARGE170. The complete knee joint model with all its components, which is meshed in ANSYS, is shown in Fig. 17.

The boundary conditions applied during the FEA simulation are shown in Table 3. For the FEA simulation the distal portion of the tibia was fixed in all degrees of freedom. The femur was fixed to allow displacement in all degrees of freedom (DOF) but no rotation. A vertical force of 1,150 N was applied on the

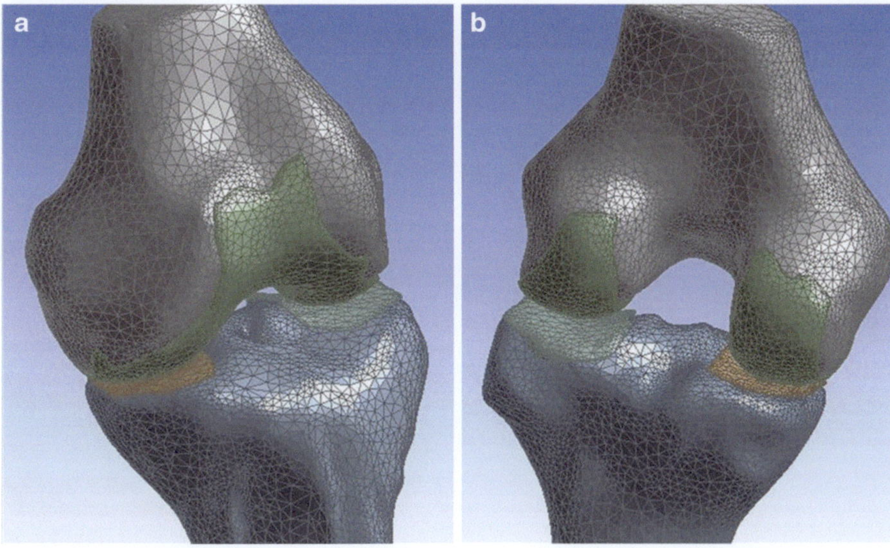

Fig. 15 (**a**) *Anterior view* and (**b**) *posterior view* of the knee joint model with all its components exported into ANSYS Workbench

Table 2 Material properties of the knee components used for finite element simulations

Component	Characteristics	Mechanical properties
Bone[a]	Elastic	$E = 400$ MPa, $\nu = 0.3$
Cartilage[b]	Elastic	$E = 15$ MPa, $\nu = 0.475$
Menisci[b]	Transversely isotropic elastic	$E_{axial/radial} = 20$ MPa, $E_{circum} = 150$ MPa, $\nu_{in\text{-}plane} = 0.2$, $\nu_{out\text{-}of\text{-}plane} = 0.3$, Shear modulus = 57.7 MPa

[a] Vadher et al. 2006
[b] Zielinska and Haut Donahue 2006

surface of the femur in the direction of the joint. This represents the force of the gait cycle for a full extension position (Peña et al. 2005; Sathasivam and Walker 1997). Although the stress distribution along all the knee components is calculated, focus was mainly on the menisci. In this regard a parametric study was performed with varying meniscus properties.

Other than the menisci with usual properties, two other extremes were also studied, one with mechanical properties ten times lower than usual (soft or compliant menisci), and one with mechanical properties ten times higher than usual (stiff menisci). In all the cases the Poisson's ratio was kept constant. Detailed material properties of the three different types of menisci considered for the parametric study are given in Table 4.

The stress distribution, strain distribution, and the total deformations of the femur and tibia with normal, stiff, and compliant menisci are depicted in Figs. 18, 19, and 20. It is found that the properties of menisci do not significantly affect the stresses and strains on femur and tibia. The total deformation with normal menisci was found to be higher within the entire medial meniscus except for the

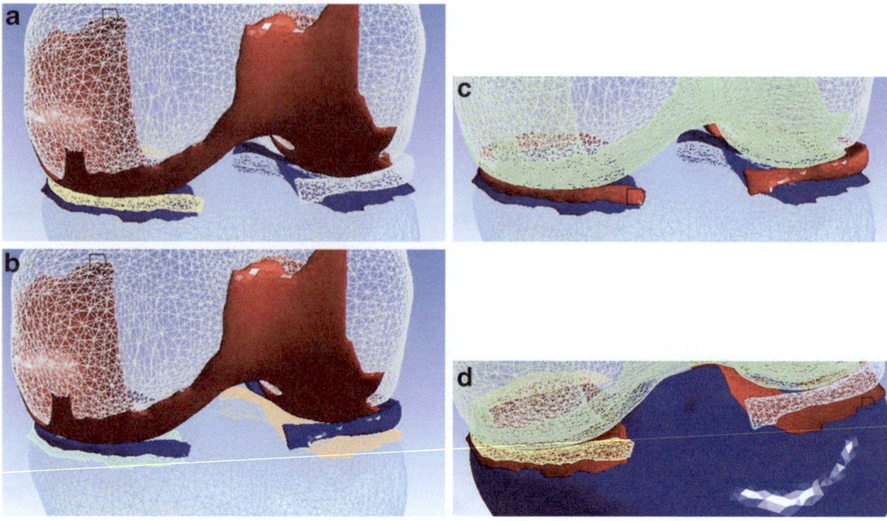

Fig. 16 Four contact pairs, which were setup in ANSYS: (**a**) femoral cartilage and tibial cartilage, (**b**) femoral cartilage and meniscus, (**c**) menisci and tibial cartilage, and (**d**) tibial cartilage and tibia

Fig. 17 Discretized knee joint model

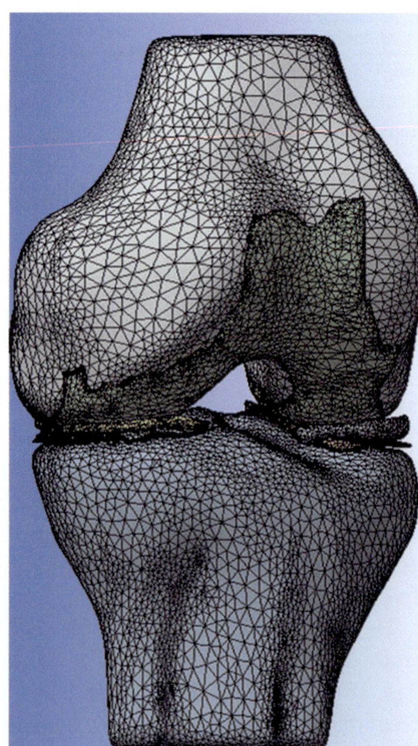

Table 3 Boundary conditions applied in finite element analysis

Type of boundary condition	Condition applied
Fixed displacement	Distal portion of Tibia fixed in all degrees of freedom
Fixed rotation	Femur with zero rotation and free displacement
Force	Axial force of 1,150 N in negative x-direction

Table 4 Material properties of the menisci used for finite element simulations

Component	Characteristics	Mechanical properties
Normal menisci[a]	Transversely isotropic elastic	$E_{axial/radial} = 20$ MPa, $E_{circum} = 150$ MPa, $\nu_{in\text{-}plane} = 0.2$, $\nu_{out\text{-}of\text{-}plane} = 0.3$, Shear modulus = 57.7 MPa
Stiff menisci	Transversely isotropic elastic	$E_{axial/radial} = 200$ MPa, $E_{circum} = 1{,}500$ MPa, $\nu_{in\text{-}plane} = 0.2$, $\nu_{out\text{-}of\text{-}plane} = 0.3$, Shear modulus = 577 MPa
Soft menisci	Transversely isotropic elastic	$E_{axial/radial} = 2$ MPa, $E_{circum} = 15$ MPa, $\nu_{in\text{-}plane} = 0.2$, $\nu_{out\text{-}of\text{-}plane} = 0.3$, Shear modulus = 5.77 MPa

[a] Zielinska and Haut Donahue (2006)

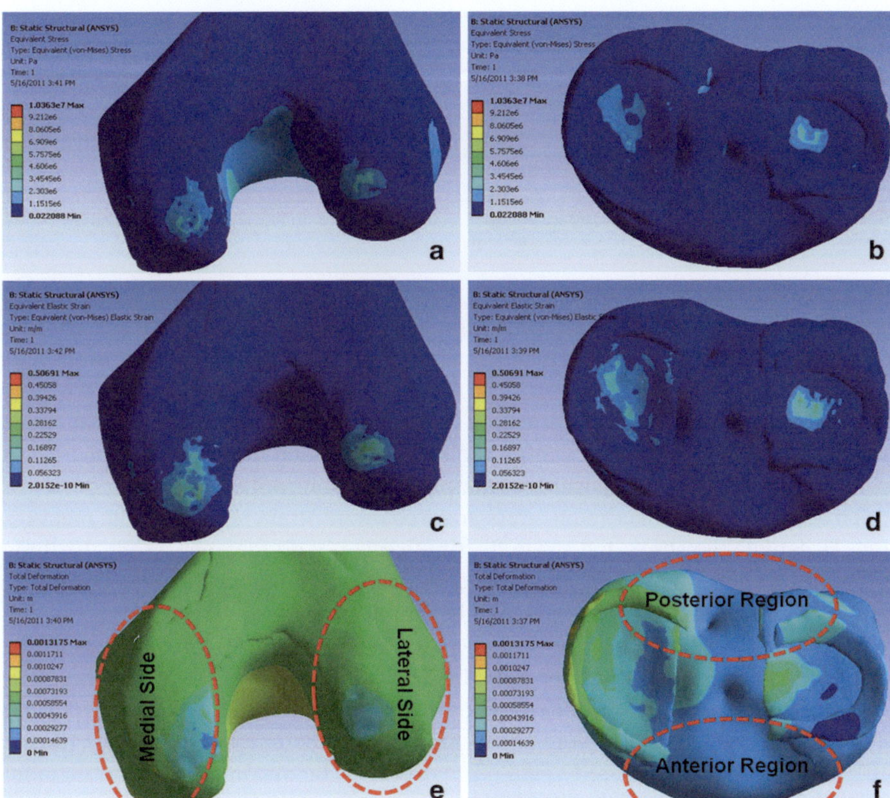

Fig. 18 Stress distribution with normal menisci (**a**) stresses in femur, (**b**) stresses in tibia, (**c**) strains in femur, (**d**) strains in tibia, (**e**) total deformation in femur, and (**f**) total deformation in tibia

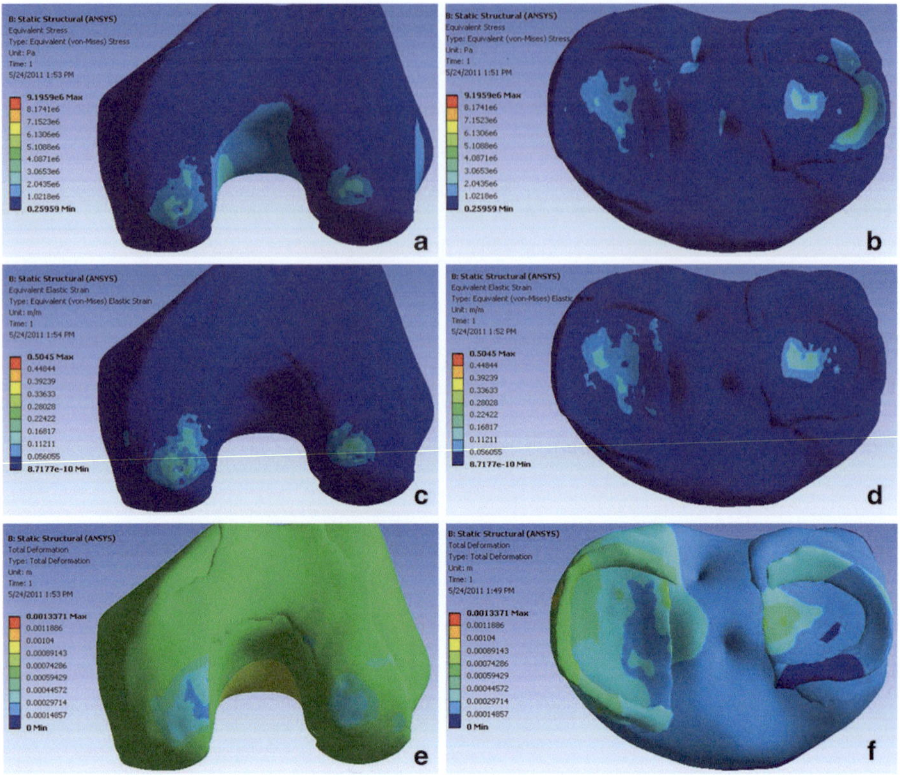

Fig. 19 Stress distribution with stiff menisci (**a**) stresses in femur, (**b**) stresses in tibia, (**c**) strains in femur, (**d**) strains in tibia, (**e**) total deformation in femur, and (**f**) total deformation in tibia

anterior portion. The total deformation with stiff menisci and soft menisci were found to be higher, within the central region of the medial meniscus and the medial meniscus on the posterior portion of the central region, respectively.

To account for the maximum stress and maximum strain regions on the menisci, Fig. 21 depicts the stress and strain distributions only on the menisci for the three different types (normal, stiff, and soft) considered in the present study. The maximum stress and maximum strain on the normal menisci were observed in the lateral meniscus on the anterior region close to the outer portion of the joint. The maximum stress was found to be 2.1 MPa. For stiff menisci maximum strain was observed in the lateral meniscus in the anterior region close to the outer portion of the joint. Higher strains were found to occur on the external periphery for the stiff menisci model compared to the other models. The maximum stress was found to be 6.32 MPa in the lateral meniscus. For soft menisci the maximum strain was observed in the lateral meniscus in the central posterior region.

It was observed that smaller strain occurred in the lateral meniscus in the posterior region close the meniscal horn attachments. The maximum stress was

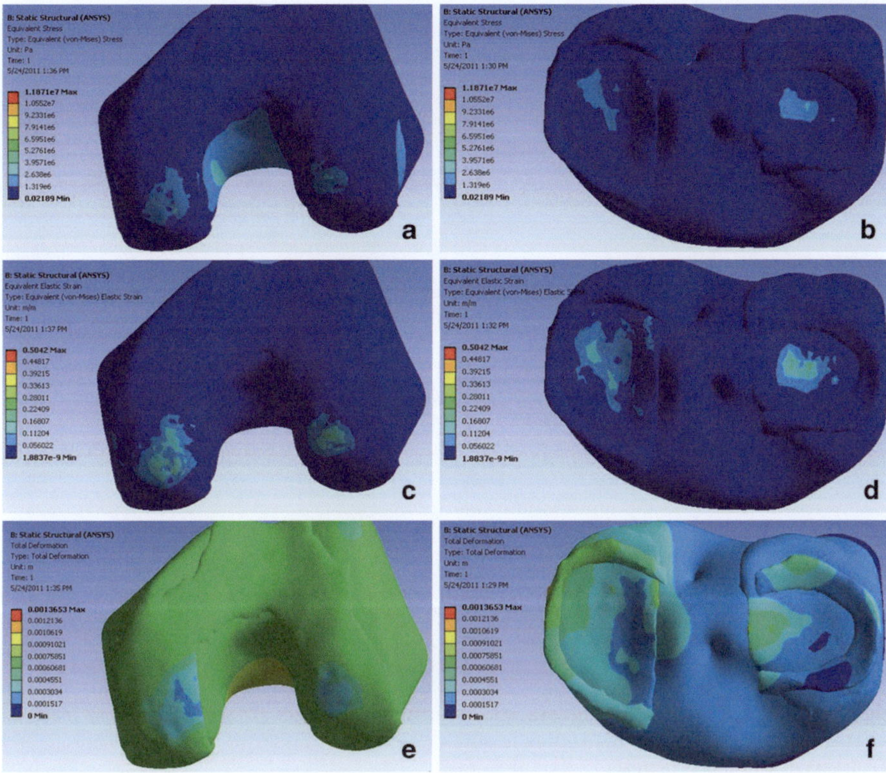

Fig. 20 Stress distribution with compliant menisci (**a**) stresses in femur, (**b**) stresses in tibia, (**c**) strains in femur, (**d**) strains in tibia, (**e**) total deformation in femur, and (**f**) total deformation in tibia

found to be 0.42 MPa in the lateral meniscus. It was also observed that the soft menisci model had more regions of high stresses compared to the other models. High stresses may lead to degradation of the meniscus or pain to the patient. This may be an early warning sign of possible osteoarthritis.

4 Conclusions

This work utilizes a geometric modeling based approach to study in-vivo contact behaviors in complex systems such as human knee joints. An approach to develop a realistic 3D model composed of entire human knee joint components including femur, tibia, cartilages, and menisci has been discussed. The construction of 3D model started with an MRI image set of human knee joint obtained using Siemens 7T scanner and then using 3D doctor software to trace out the boundaries of all the

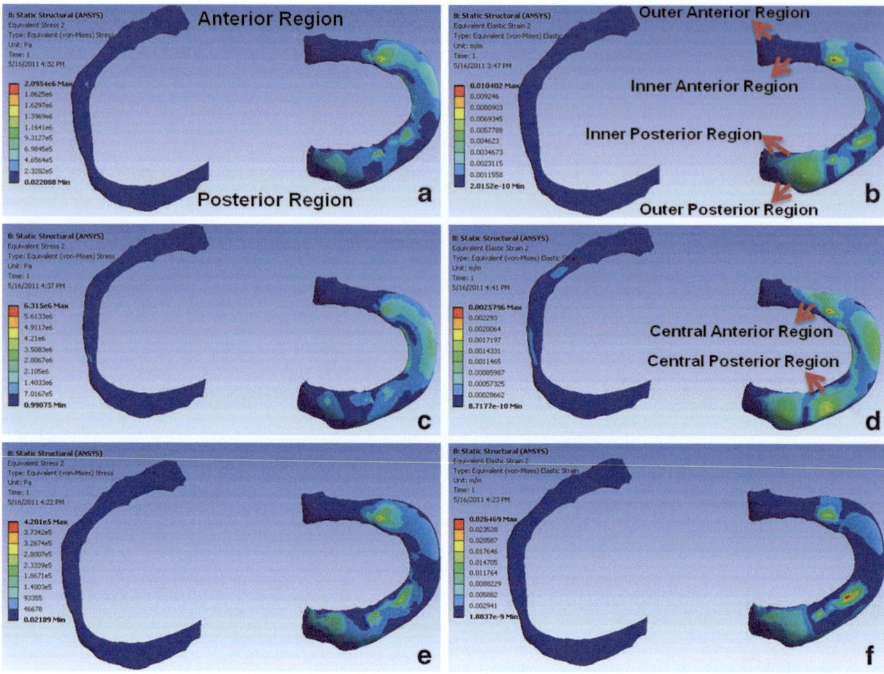

Fig. 21 Stress distribution in menisci (**a**) stresses in normal menisci, (**b**) strains in normal menisci, (**c**) stresses in stiff menisci, (**d**) strains in stiff menisci, (**e**) stresses in compliant menisci, and (**f**) strains in compliant menisci

components. To ease out the complexities due to the tracing of the components, smoothing operations were performed. The constructed model was then imported into ANSYS Workbench for FEA simulations. To study the load bearing performance of menisci a parametric study was performed to analyze the effect of damaged menisci or meniscal implants. The parametric study involved manipulation of mechanical properties of menisci by either ten times higher (hard menisci) or ten times lower (soft menisci) than the usual properties available in literature. The locations of maximum stress and strain were dependent on the mechanical properties of menisci. It is found that there is a significant dependence of maximum stress values on the mechanical properties of menisci. For example, compared to a baseline case, the maximum stress value was three times higher when mechanical properties of menisci increased by ten times and five times lower when mechanical properties of menisci decreased by ten times. It is also observed in the case of compliant menisci a larger region within the menisci was subjected to high stresses. Such a case causes a concern of premature onset of osteoarthritis. A comprehensive and detailed understanding of the contact behaviors of menisci is vital for engineering of meniscal implants.

Acknowledgments Dr. Peter Walker, Miriam Chaudhary, and Sally Arno at the Laboratory for Minimally Invasive Surgery of the Department of Orthopedic Surgery, NYU Hospital for Joint Diseases are acknowledged for the help with specimen preparations and mechanical testing. The authors acknowledge Dr. Ravinder R. Regatte at NYU Langone Medical Center for the help with MRI imaging and technical discussions. Ansys Inc is thanked for providing reduced price academic research licenses. The authors acknowledge the NYU – NYU-Poly joint seed grant and the MAE department at NYU-Poly for the facilities and support provided.

References

Alatakis S, Naidoo P (2009) MR imaging of meniscal and cartilage injuries of the knee. Magn Reson Imaging Clin North Am 17:741–756

Arokoski JPA et al (2000) Normal and pathological adaptations of articular cartilage to joint loading. Scand J Med Sci Sports 10:186–198

Baldwin MA et al (2009) Efficient probabilistic representation of tibiofemoral soft tissue constraint. Comput Methods Biomech Biomed Engin 12:651–659

Baldwin MA et al (2010) Development of subject-specific and statistical shape models of the knee using an efficient segmentation and mesh-morphing approach. Comput Methods Programs Biomed 97:232–240

Bendjaballah MZ, Shirazi-Adl A, Zukor DJ (1995) Biomechanics of the human knee joint in compression: reconstruction, mesh generation and finite element analysis. Knee 2:69–79

Bendjaballah MZ, Shirazi-Adl A, Zukor DJ (1997) Finite element analysis of human knee joint in varus-valgus. Clin Biomech 12:139–148

Blankevoort L et al (1991) Articular contact in a three-dimensional model of the knee. J Biomech 24:1019–1031

Busse H et al (2008) In vivo glenohumeral analysis using 3D MRI models and a flexible software tool: feasibility and precision. J Magn Reson Imaging 27:162–170

Chang G et al (2010) Biochemical and physiological MR imaging of skeletal muscle at 7 tesla and above. Semin Musculoskelet Radiol 14:269–278

Chaudhary ME, John D, Walker PS (2011) Relative stability provided by the medial meniscus and cruciate ligaments at high and low axial compressions. In: ASME summer bioengineering conference, Farmington, PA, USA, 22–25 June

Cheng YJ et al (2010a) Prevalence of doctor-diagnosed arthritis and arthritis-attributable activity limitation-United States, 2007–2009. JAMA 304:2238–2239

Cheng YY et al (2010b) 3D FEA of high-performance polyethylene fiber reinforced maxillary dentures. Dent Mater 26:e211–e219

Cohen ZA et al (1999) Knee cartilage topography, thickness, and contact areas from MRI: in-vitro calibration and in-vivo measurements. Osteoarthritis Cartilage 7:95–109

DeFrate LE et al (2004) In vivo tibiofemoral contact analysis using 3D MRI-based knee models. J Biomech 37:1499–1504

Englund M (2008) The role of the meniscus in osteoarthritis genesis. Rheum Dis Clin North Am 34:573–579

Englund M, Guermazi A, Lohmander LS (2009) The meniscus in knee osteoarthritis. Rheum Dis Clin North Am 35:579–590

Fan W et al (2008) Identification of a C-shaped canal system in mandibular second molars–Part III: anatomic features revealed by digital subtraction radiography. J Endod 34:1187–1190

Ferraz EG et al (2011) Effect of different surface processing protocols in three-dimensional images for rapid prototyping. Adv Eng Softw 142:332–335

Garstang SV, Stitik TP (2006) Osteoarthritis: epidemiology, risk factors, and pathophysiology. Am J Phys Med Rehabil 85:S2–S11

Gíslason MK, Stansfield B, Nash DH (2010) Finite element model creation and stability considerations of complex biological articulation: the human wrist joint. Med Eng Phys 32:523–531

Guo Y, Zhang X, Chen W (2009) Three-dimensional finite element simulation of total knee joint in gait cycle. Acta Mech Solida Sin 22:347–351

Hao Z et al (2007) A finite element 3D model of in vivo human knee joint based on MRI for the tibiofemoral joint contact analysis. In: 1st international conference on digital human modeling, Beijing, China, Springer, Berlin/Heidelberg, LNCS, vol 4561, pp 616–622

Haut Donahue TL et al (2002) A finite element model of the human knee joint for the study of tibio-femoral contact. J Biomech Eng 124:273–280

Haut Donahue TL et al (2003) How the stiffness of meniscal attachments and meniscal material properties affect tibio-femoral contact pressure computed using a validated finite element model of the human knee joint. J Biomech 36:19–34

Hootman JM, Helmick CG (2006) Projections of US prevalence of arthritis and associated activity limitations. Arthritis Rheum 54:226–229

Liu P (2011) Construction of 3D finite element model of upper end of tibia. In: 3rd international conference on measuring technology and mechatronics automation, Shanghai, 6–7 Jan, pp 766–769

Loeser RF, Shakoor N (2003) Aging or osteoarthritis: which is the problem? Rheum Dis Clin North Am 29:653–673

Madelin G et al (2010) Sodium inversion recovery MRI of the knee joint in vivo at 7T. J Magn Reson 207:42–52

Magne P (2007) Efficient 3D finite element analysis of dental restorative procedures using micro-CT data. Dent Mater 23:539–548

Ni Y et al (2008) Quantitative morphology of facial nerve based on three-dimensional reconstruction of temporal bone. Otolaryngol Head Neck Surg 138:23–29

Pakin SK et al (2006) Ultra-high-field MRI of knee joint at 7.0T: preliminary experience. Acad Radiol 13:1135–1142

Peña E et al (2005) Finite element analysis of the effect of meniscal tears and meniscectomies on human knee biomechanics. Clin Biomech 20:498–507

Peña E et al (2006) Why lateral meniscectomy is more dangerous than medial meniscectomy. A finite element study. J Orthop Res 24:1001–1010

Peña E et al (2008) Computer simulation of damage on distal femoral articular cartilage after meniscectomies. Comput Biol Med 38:69–81

Périé D, Hobatho MC (1998) In vivo determination of contact areas and pressure of the femorotibial joint using non-linear finite element analysis. Clin Biomech 13:394–402

Regatte RR, Schweitzer ME (2007) Ultra-high-field MRI of the musculoskeletal system at 7.0T. J Magn Reson Imaging 25:262–269

Regatte RR, Schweitzer ME (2008) Novel contrast mechanisms at 3 tesla and 7 tesla. Semin Musculoskelet Radiol 12:266–280

Rho JY et al (2002) Microstructural elasticity and regional heterogeneity in human femoral bone of various ages examined by nano-indentation. J Biomech 35:189–198

Robertson DD et al (2003) Knee joint biomechanics: relevance to imaging. Semin Musculoskelet Radiol 7:43–58

Sathasivam S, Walker PS (1997) A computer model with surface friction for the prediction of total knee kinematics. J Biomech 30:177–184

Sharma L et al (2001) The role of knee alignment in disease progression and functional decline in knee osteoarthritis. JAMA 286:188–195

Shin DS, Lee K, Kim D (2007) Biomechanical study of lumbar spine with dynamic stabilization device using finite element method. Comput Aided Des 39:559–567

Shirazi R, Shirazi-Adl A (2009) Analysis of partial meniscectomy and ACL reconstruction in knee joint biomechanics under a combined loading. Clin Biomech 24:755–761

Song Y et al (2008) Meniscectomy alters the dynamic deformational behavior and cumulative strain of tibial articular cartilage in knee joints subjected to cyclic loads. Osteoarthritis Cartilage 16:1545–1554

Tang S-T et al (2009) Clinical value of pelvic 3-dimensional magnetic resonance image reconstruction in anorectal malformations. J Pediatr Surg 44:2369–2374

Trilha M Jr et al (2009) Simulação numérica tridimensional da mecânica do joelho humano. Acta Ortop Bras 17:18–23

Vadher SP et al (2006) Finite element modeling following partial meniscectomy: effect of various size of resection. In: 28th annual international conference of the IEEE "Engineering in Medicine and Biology Society", New York

Vaziri A et al (2008) Influence of meniscectomy and meniscus replacement on the stress distribution in human knee joint. Ann Biomed Eng 36:1335–1344

Vedi V et al (1999) Meniscal movement: an in-vivo study using dynamic MRI. J Bone Joint Surg Br 81-B:37–41

Walker PSA et al (1997) Knee simulating machine for performance evaluation of total knee replacements. J Biomech 30:83–89

Walker PS et al (2009) Design features of total knees for achieving normal knee motion characteristics. J Arthroplasty 24:475–483

Walker PS et al (2011) Preclinical evaluation method for total knees designed to restore normal knee mechanics. J Arthroplasty 26:152–160

Yang NH, Nayeb-Hashemi H, Canavan PK (2007) The effects of tibiofemoral angle and body weight on the stress field in the knee joint. In: International mechanical engineering Congress and exposition, Seattle, WA, USA, 11–15 Nov

Yang NH et al (2010) Effect of frontal plane tibiofemoral angle on the stress and strain at the knee cartilage during the stance phase of gait. J Orthop Res 28:1539–1547

Zielinska B, Haut Donahue TL (2006) 3D finite element model of meniscectomy: changes in joint contact behavior. J Biomech Eng 128:115–123

Hip Prostheses Computational Modeling: FEM Simulations Integrated with Fatigue Mechanical Tests

Ilaria Campioni, Gianluca Notarangelo, Ugo Andreaus, Angelo Ventura, and Claudia Giacomozzi

Abstract Hip prosthetic implants represent a consolidated and successful solution to restore functional gait in patients affected by a wide range of disease including osteoarthritis degeneration processes, cancer effects, osteoporosis, traumatic injuries. While at the beginning of the worldwide dissemination of this joint replacement methodology the target patients were mainly part of a quite old population, characterized by moderate physical activity and only asking for restoring of an acceptable quality of life and of functional gait and standing, the scenario significantly changed along the years. Nowadays, hip implants are also used to treat young and active patients; moreover, the amount of expected life years is more and more increasing; and, last but not least, the average body mass of western populations is increasing as well. As an overall consequence of the above changes, hip implants need to be highly performing: they have to cope with many years of repetitive high stresses not only due to regular locomotor daily activities but also to sports, excessive loading, wear and ageing effects. It is thus of extremely relevance for Researchers, Industry, Health Authorities and Users, to gain deeper and deeper knowledge of mechanical properties and performance of hip prosthesis components, and behavior of the musculo-skeletal system which hosts the implant. Potentially dangerous conditions should be clearly identified and investigated so as to prevent implant structural failure, being implant revision highly demanding for patients and health structures in terms of worsening of health status and increase of

I. Campioni • G. Notarangelo • C. Giacomozzi (✉)
Department of Technology and Health, Italian National Institute of Health (ISS), Rome, Italy
e-mail: claudia.giacomozzi@iss.it

U. Andreaus
Department of Structural and Geotechnical Engineering, Faculty of Civil and Industrial Engineering, Sapienza University of Rome, Rome, Italy

A. Ventura
Department of Orthopedics (II Division), C.T.O. (Orthopedic Trauma Center) "A. Alesini", Rome, Italy

D. Iacoviello and U. Andreaus (eds.), *Biomedical Imaging and Computational Modeling in Biomechanics*, Lecture Notes in Computational Vision and Biomechanics 4,
DOI 10.1007/978-94-007-4270-3_5,

overall assistance costs. The present chapter investigates the potentialities of research studies in the field of hip implants biomechanics which rely on a synergy between FE modeling and experimental mechanical fatigue tests and whose main goal is to infer about related risks. An example of the implemented methodology is described, and few practical applications are reported, analyzed and commented from clinical, biomechanical and regulatory point of view.

Keywords Hip prosthesis • Mechanical testing • Fatigue • Finite Element Method (FEM) • Finite Element Analysis (FEA) • ISO Standard

1 Introduction

Nowadays the number of hip arthroplasty interventions is still rising at a worldwide level, especially among quite young and demanding patients. The exact amount of interventions is not easy to be found for all Countries. Numbers can be found on the web with respect to U.S., where the incidence of hip replacement was 0.04% in 2004 (US Census Bureau, Population Estimates, 2004), or Europe, where incidence was about 0.1% in 2005, up to 0.16% if only considering Countries of Western Europe (Kiefer 2007). More specifically, an Italian HTA Report published in 2008 reported that incidence in Italy was 0.09% with an increase of 32% in the number of total hip replacements from 1999 to 2005 (Age.na.s HTA Report 2008). Advances in stem design, material properties, cementing techniques and surgical skills have significantly increased the lifetime of implants, thus the hip prosthesis still represents one of the greatest successes of modern orthopaedic surgery. However, structural failure or mobilization are still occurring: again, numbers are not easily found; information mainly come from National Registries for joint replacement which, however, are not established in each Country and, even more important, are sometimes compiled on a voluntary basis. Few interesting examples: (i) data from the Swedish joint Registry, which was established in the 1970s, showed that Swedish revision rate for hip surgery reduced to 8% in 2000, in comparison with a reported revision rate of 18% in U.S. for the same year (Herberts and Malchau 2000; Lie et al. 2004); (ii) UK National Joint Registry, the largest joint Registry in the world in 2010, reported a 12–15% of revision rate at 5 years (Hodgkinson et al. 2011); (iii) in 2004, the Norwegian Arthroplasty Register proved that 11% of all primary prostheses failed within 10 years (Lie et al. 2004). Even though not representative of the current worldwide scenarios, the above examples well explain the need for investigation which go deeper into the implant biomechanics.

Basically, the purpose of an artificial hip joint is to create a system which restores as much as possible the kinematics of the locomotor system and allows it to withstand physiological loads while minimizing wear and friction and avoiding harmful reactions in the body. Usually, possible hip implant failure mechanisms are classified on the basis of the functional duration the prosthesis reaches once implanted: (i) short term failures, such as infection or dislocation; (ii) mean term failures, mainly due to mechanical failure of one or more components of the implant, good examples being the loosening of a component due to structural failure of the embedding medium, or the stress shielding due to severe atrophy of the periprosthetic bone tissue; (iii) long term failure due to fatigue failure of one or more components. The interest of the present study is particularly addressed to this last aspect, which limits the average functional duration of the implant and which is a major cause for surgical revision. More specifically, the study aims at understanding which potential failures are strictly due to mechanical properties of the implant, and which might be due to concurrent adverse mechanical actions or implantation conditions.

The study addresses the investigation of the above issues by developing a Finite Element (FE) modeling -based methodology and by performing proper experimental fatigue tests. The latter are implemented in a laboratory context, according to Technical International Standards voluntarily applied by both Manufacturers and Notified Bodies to assess the compliance of Medical Devices to specific regulations before they are allowed to enter the market.

Despite the relevance of standardized experimental fatigue testing, however, only few studies have been found up to now in the peer-reviewed International literature which deal with this issue. In the interest of the present chapter, it is worth to briefly discuss them here below.

Raimondi and Pietrabissa (1999), investigated the appropriateness of the International Standard ISO 7206 to assess the fatigue reliability of hip prostheses with special attention to femoral stems. They used numerical simulation – software application Abaqus 5.7; titanium alloy stem model with 9,000 brick elements – to reproduce test conditions as defined in the above Standard and to vary the input parameters. More specifically, they simulated the standard test condition – abduction angle $\alpha = 10°$, flexion angle $\beta = 9°$, constraint level d $= 80$ mm, maximum load 2,300 N – plus a number of different conditions where, being all other test parameters fixed, only one parameter changed, namely: α also fixed at $9°30'$ and $10°30'$; β also fixed at $8°$ and $10°$, constraint level d also fixed at 78 and 82 mm; six types of stem distal constrains. As output parameter they focused on the static maximum stem stress, considering it as directly related to the expected fatigue lifetime of the stem. The highest stress (464 MPa) was found in correspondence with a standard positioning without cement layer and with all stem nodes fixed at upper surface of a virtual cement layer to block the stem (maximum stress under standard conditions and thin cement layer was 310 MPa). In their study, however, no experimental tests were implemented to validate the results from the computational models.

Chao (2008) conducted quite a different investigation, based on analytical computation rather than on numerical simulation, which posed a very important question about the suitability of the ISO 7206-4: 1989 indications for femoral stem fatigue assessment, more specifically with respect to the loading indications. His investigation highlighted that, for the specific titanium alloy stem under examination, tests performed under the ISO 7206-4: 1989 conditions delivered ratios of maximum stress by endurance limits equal to 0.876 at neck and 0.384 at the stem constraint. In order to predict a mechanical failure, the above ratio should be equal or greater than 1, thus under the test condition the stem was proved suitable for the intended use (even though 0.876 is indeed scarcely conservative when compared with the threshold risk of 1). However, when a higher load is applied, which is reasonable for heavier patients, things change dramatically: as an example, for a patient's weight of 1,000 N, and an average hip loading of 4,000 N, the two above ratios become 1.524 and 0.669 respectively, thus showing the unsuitability of the standard indications in case of obese patients.

Griza et al. (2009) conducted a thorough investigation on a specific cemented femoral stem whose failure had happened after 5.5 years from the implant. The investigation included failure and metallographic analysis, the setup of numerical simulation to detect values and location of static maximum stress, and the implementation of fatigue tests to determine the stress *vs* number of cycles curve. Relevant details of the study were: analysis applied to a steel alloy femoral stem; FE software application Abaqus 6.5; 13,511 stem elements in total, both hexahedral and tetrahedral, parametric constraint level as indicated by the ISO 7206-4: 2002, applied maximum load 2,300 N. Results from numerical simulation were well in agreement with the failure analysis and the prediction of the stem lifetime: von Mises stress was found equal to 210 and 235 MPa at the constraint level – laterally and medially respectively; failure number of cycles was estimated around 750,000 on the basis of ASTM F-745. The study thus proved that a thorough mechanical assessment of the stem before implantation might have allowed to predict its failure soon after loosening.

Moreover, Ploeg et al. (2009) addressed the relevant issue of hip stem fatigue test prediction methods based on linear elastic finite element analysis, cyclic stress-strain curves of the material, and specific correction methods. Despite some limitations in the study, which was performed under simplified conditions – i.e. axial loading configuration without torsion, regular and symmetric stem geometry – they highlighted the potentiality of some prediction methods. Interestingly, they found that at least for the titanium alloys included in their study, the classical S-N method under-estimated the stem life. While indeed interesting and innovative, this complex study remains of difficult applicability in all those cases when there is the need to cope with several different stem materials and geometries.

Main aim of the present study was rather to validate and implement computational models and static stress analysis in order to identify critical test configurations, which might lead to fatigue failure of the prosthetic implant. Then, particularly critical conditions have been tested mechanically through standard fatigue test. To perform the study, proper finite element models have been developed starting from the ISO 7206-4: 2002 operating indications and a set of simulations have been implemented by only varying one parameter per test, i.e. different angular positions on both frontal and sagittal plane, step increase of maximum loading, different constraint levels. Ranges of variation for the parameters were established by discussing implantation issues from a surgical point of view, thus taking into account those conditions which might occur in a clinical intervention. Distribution of maximum stress was then analyzed and discussed with respect to the effect on the stem fatigue behavior. More specifically, two main models were created: (i) Model1, fully compliant with the mechanical test configuration in ISO 7206-4

(ISO 7206-4: 2002) and 8 (ISO 7206-8: 1995): the model was used to identify the stem most critical orientation as for abduction and flexion angles; (ii) Model2, which is based on the modeling of the human femur with prosthesis inside; this second model, whose configuration was quite similar to the real clinical one, was used to better understand the modification of stress distribution due to the stem most critical orientation in the two planes. A mechanical fatigue test was also performed under the most critical condition among those investigated through simulations: the test allowed to validate the numerical model and to highlight the potential risk of failure of a misaligned implant.

At the level of the EU Market for Medical Devices, a relevant change occurred in 2010, when the section 4 of ISO 7206 was revised and the level of stem constraint was brought back to the level of 80 mm, as it was in the 1998 revision. The shift of the constraint level along the axis renders mechanical fatigue tests significantly more critical for smaller stems. This issue has been further investigated at the end of the present study, and a brief discussion is reported on the basis of some preliminary simulations.

2 Materials and Methods

2.1 *Fatigue Experimental Tests*

The fatigue behavior of prosthetic femoral stems, in Europe classified as class III Medical Devices, are commonly analyzed by means of experimental tests. These tests are usually performed according to Technical International Standards voluntarily applied by both Manufacturers and Notified Bodies. The fatigue test is one of the most common laboratory tests to evaluate the reliability of a prosthesis as underlined in (Paul 1997); at a European level it is performed according to the ISO 7206 standard, which describes the test apparatus and procedure to assess the fatigue resistance of femoral stems principally in two different configurations, the first configuration with the stem embedded in a cement mantle up to a pre-defined distance from the centre of the head (Raimondi and Pietrabissa 1999; Griza et al. 2009), the second one with the stem cemented up to the osteotomy plane (Griza et al. 2008b).

In particular, parts 4, 6 and 8 of the ISO Standard 7206 are taken into account in the implementation of the fatigue experimental tests. In the ISO 7206-4 "Determination of Endurance Properties of Stemmed Femoral Components", revision of 2002, indications are given to constrain the stem in the embedding medium with a distance D from the centre of the head and with a proper orientation with respect to adduction (angle α) and flexion (angle β) as shown in Fig. 1.

Being CT the distance between the femoral head centre and the most distal point of the stem, the constraint distance D (potting level) is obtained by the product 0.4*CT if CT $\leq$ 200 mm and by the difference CT-100 for CT greater than 200 mm. Angles are set to 10° (angle α) and to 9° (angle β) for CT $\leq$ 200 mm, and 0° and 4° respectively

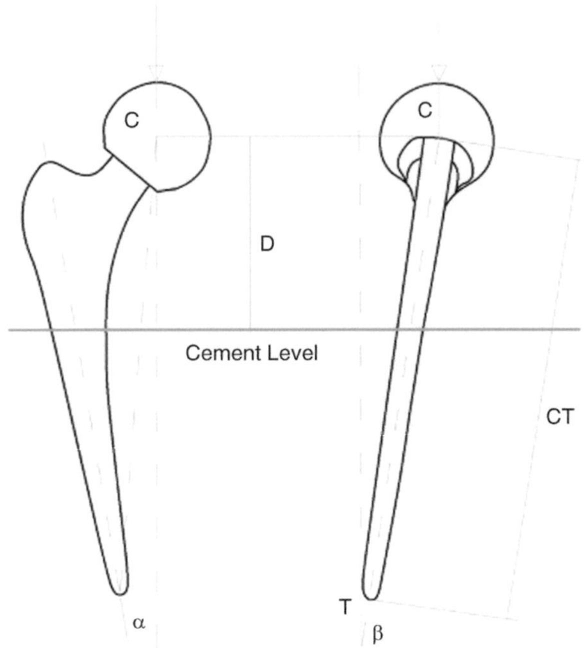

Fig. 1 Orientation of the stem for fatigue test (ISO 7206-4: 2002)

for CT greater than 200 mm. The load to be applied must be sinusoidal, ranging from 300 to 2,300 N with a frequency selectable in between 1 and 30 Hz. The fatigue test has to be considered terminated when one of the following events occurs: the deflection of the head exceeds 5 mm, fracture of the specimen occurs, or the maximum number of requested cycles (5×10^6 cycles) is reached.

In the present study, mechanical tests are performed on a MTS 858 Table Top System test machine (MTS, MN, USA) with a load cell of 25 kN. The system has appropriate performance characteristics which are in compliance with those indicated in ISO standard (dynamic loading waveform sinusoidal at the primary frequency, error in applied load not greater than $\pm 2\%$ at maximum load, accuracy of 0.5 mm to monitor the deflection of the head of the test specimen). In particular, as demonstrated by the results obtained in some implemented tests described in Giacomozzi et al. (2010), the test machine equipped with the requested experimental chain is able to apply a sinusoidal load with an error less than $\pm 1\%$ at maximum value and to monitor displacement with an accuracy of 0.01 mm.

2.2 *Finite Element Simulation: Modell*

Finite Element models used in this study to carry out the finite element analysis with the software COMSOL Multiphysics (COMSOL Multiphysics, COMSOL AB, Sweden), were created by digital acquisition of 3D model of femoral stem with 3D-

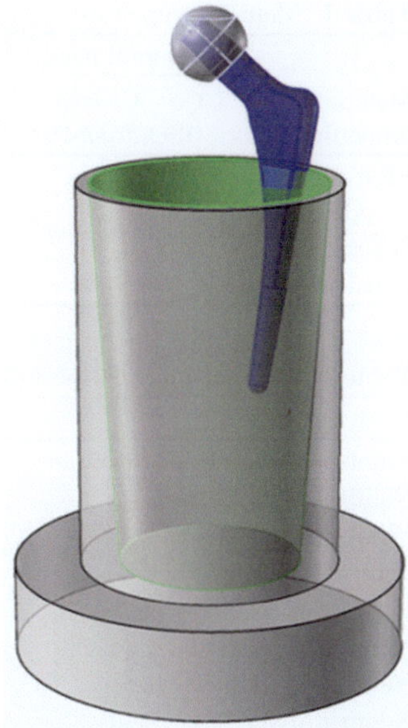

Fig. 2 Complete model created to implement finite element analysis: Model1

laser scanner (SCANNY3D- base, SCANNY3D s.r.l., Italy). A post-elaboration was then performed with the software CAD Rhinoceros (Rhinoceros 4.0, McNeel, USA) to improve scanned surfaces and to refine the model (Fig. 2).

The femoral stem adopted is a straight stem (size: 115 mm) for an uncemented implant, made of titanium alloy, with a neck-shaft angle of 125° and cone 12/14.

The aim of the numerical model was to reproduce the real configuration used during mechanical tests performed in the laboratory. Thus, the complete model consisted of: head and stem of hip prosthesis; cement mantle; an external metallic fixture dimensionally equal to the real specimen holder.

In order to respect the boundary conditions required in the Standard, the prosthesis was enveloped with a cement mantle inside the metallic fixture and then rigidly fixed at 0.4CT. A load of 3 kN was applied at the centre of the head on a surface of 311 mm^2, that has the same dimension of mechanical element used in the laboratory tests for heads of 28 mm of diameter, so as to distribute the stress over the prosthesis head. Finally, the base of the metallic fixture was bound.

As for the interfaces of the model: prosthesis/cement and cement/ metallic fixture interfaces were assumed to be fully bonded. The first assumption, which is the most appropriate interface boundary condition to be used especially when the cement mantle is quite thick, was done in agreement with Raimondi and Pietrabissa (1999) and Griza et al. (2008a).

As for the stem orientation, and in order to investigate how different testing conditions are associated with the variation of stem maximum stress, not only the standard configuration was implemented, with stem adduction angle $\alpha = 10°$ and stem flexion angle $\beta = 9°$, but also other angular positions were implemented and examined, with angle α ranging from 2° to 15° and angle β ranging from 2° to 13°. Finally, the mechanical properties of the materials used in the simulation are reported in Table 1.

Table 1 Material properties

	Femoral head	Femoral stem	Cement	Fixture	Femur
Material properties	Co – Cr – Mo (ISO 5832-IV)	Ti – 6Al – 4V (ISO 5832-III)	PMMA (Technovit 4071)	Stainless steel	Cortical bone
Elastic modulus [GPa]	220	114	3	200	17
Poisson's ratio	0.33	0.24	0.30	0.33	0.29
Density [kg/m^3]	8,830	4,400	1,190	7,850	2,000

Table 2 Characteristics of mesh for Model1 and Model2

Mesh parameters	Model1	Model2
Number of degrees of freedom	102,666	123,531
Number of elements	23,834	28,150
Number of mesh points	4,633	5,649
Tetrahedral	23,834	28,150
Number of boundary elements	7,307	7,174
Number of triangular elements	4,485	7,174
Number of edge elements	543	813
Number of vertex elements	50	84
Minimum element quality (value between 0 and 1)	0.245	0.258

Regarding the procedure for the mesh generation of the model, special attention was paid to the surface of the stem at the level of its connection with the cement mantle. A mesh refinement was employed by setting some boundary mesh parameters: element growth rate at 1.4, mesh curvature factor at 0.4 and mesh curvature cutoff at 0.02. The mesh characteristics of the complete model consist of 23,834 tetrahedral elements and 102,666 degrees of freedom as shown in detail in Table 2.

2.2.1 Model1: Validation of the Model

In order to validate the numerical methodology used in the present investigation, the numerical model of Griza et al. (2008a) was replicated as a first step, and the results compared. Griza's model, in fact, represents a valuable contribution in stem numerical simulation, and the cited paper contains all the relevant information about the model they created and the main parameters they included in their simulations. The aim of their work was to show how a specific cemented hip prosthesis design, created to provide axial stability, might be vulnerable to fatigue fracture. To this aim, they performed FE analysis to examine the effect of the incorporated geometrical feature on the prosthesis mechanical behavior under physiological loading; the computational algorithm was based on the ISO Standard ISO 7206-4: 2002, and simulations were implemented with the software Abaqus 6.5. As for material properties, the stem was modeled in stainless steel ASTM F-745. In the validation step of the present study, a CAD model was constructed with the same Griza's dimensions and used under the same Griza's simulation conditions. The results obtained during the validation phase were in agreement with Griza's main results: for principal stress distribution, Griza et al. reported a maximum value of 329 MPa located in correspondence with the upper surface of the cement, whilst the present validation study delivered a maximum value of 332 MPa in the same region with a difference of only 0.91%.

Another relevant issue was deeply investigated in the validation phase, i.e. the suitability of conducting numerical analysis under a *static regimen* condition, without taking into account frequency. Even though this represents a common approach in the relevant literature, the validity of such an assumption was here investigated by implementing some numerical simulations and analyzing the vibration modes of the complete model. From the analysis four vibration modes were identified, and the first frequency oscillation was found equal to 431.2 Hz, which is well above the 10 Hz frequency used in laboratory fatigue test. Furthermore, stress values in static than dynamic regimen resulted to be very similar. Based on the above findings, frequency was not taken into account in the numerical analysis conducted in the present study.

2.2.2 Model1: Method to Determine Critical Conditions

As a first operative step, efforts were made on the implementation of a model to reproduce the same conditions of the mechanical experimental tests performed in the Authors' Laboratory according with ISO Standard 7206, and also to identify the most critical conditions for femoral stem during such a test. To this second aim, a set of numerical simulations was conducted by varying the stem orientation in both the sagittal and frontal anatomical planes within angular ranges which might be reasonable both from a clinical and experimental point of view (the latter, in fact, has to cope with geometrical constraints and dimensional limitations). In particular, the adduction angle α was varied from 2° to 15° – step 1° – while the flexion angle β was kept fixed at 9° as indicated by the above ISO Standard; successively, the flexion angle β was increased from 2° to 13° – step 1° – while the adduction angle α was kept fixed at 10° as indicated in the Standard. Among the outcomes of the simulations, attention was focused on the maximum value of Von Mises stress in order to identify the stem most critical condition. The assumption was here to consider as a critical configuration the combination of the alpha and beta angles in correspondence of which the maximum value of Von Mises stress had been localized on the stem, without considering all possible angular combinations and/or smaller angular steps. Implementation of such investigations were behind the purpose of the current study; however, they are discussed in the final part of the present chapter. Besides Von Mises maximum stress, other parameters were also monitored in the simulations, i.e. first principal stress, normal stress-z direction and displacement in z direction, but they were not taken into account for the identification of the critical configuration.

2.3 Finite Element Simulation: Model2

The configuration of the Model1 as described in the above Paragraphs is different from those observed in clinical practice at the time of implantation, i.e. Model1 put the level of constraint at a certain distance along the stem, whilst stem implantation requires stem fixation up to the neck, in correspondence with the osteotomy plane. While Model1 implementation becomes relevant once a first mobilization of the implant has occurred, a more "realistic" model might help studying the primary stability of a stem or to investigate micro-motions at bone-stem or stem-cement interfaces and to obtain a true preclinical validation of the prosthetic implant (Viceconti et al. 2009: the paper reports on an example of the principal steps of a pre-clinical validation of an epiphyseal stem).

Actually, there are many studies in the literature describing the implementation of more realistic models reproducing a stem prosthesis inside the geometry of a femur, with or without cement mantle interfaces. These studies differ among them for the use of different technical reconstruction of the geometry – CT images, CAD design – , for modeling the complete femur or only the proximal part, for different boundary conditions used – simulation of muscles forces or not, contact among interfaces, high mesh resolution – or because they only focus on the investigation of a particular aspect of modeling – i.e. adequate loading conditions to represent in vivo situation, or the effects of different bonding conditions – . With respect to this last aspect, some of these studies are briefly described here below.

Mann et al. (1995) performed a study to determine whether realistic stem-cement interface characteristics might result in high cement stresses when compared to a bonded stem-cement interface and to determine if stem design parameters could be chosen to reduce peak cement stresses. To this aim, the Authors considered a 3D finite element model of cemented femoral hip components: the representation of the bone was developed from CT scan data of a "normal" proximal femur; an idealized 3 mm constant-thickness cement mantle was added around the stem, with bonded or realistic Coulomb friction stem-cement interfaces. Four stem cross-sectional geometries of titanium alloy or cobalt chromium alloy were considered. Models were represented with 1,128 elements (20-noded brick elements) and materials properties for the stem and cement were assumed to be isotropic and homogeneous. Two loading cases were considered: the first represents an estimate of loading in an active individual during stance phase of gait; the second represents a patient's postoperative loading during gait. Loads were applied to the head centre and greater trochanter. An interesting result of this study was that the maximum principal stresses predicted in the stem for the bonded and Coulomb friction interface cases were similar in magnitude and distribution (208 MPa in the bounded models and 220 MPa in friction interface models), whilst in the proximal cement mantle tensile stresses were greater than the fatigue strength for the friction interface case (10.8 MPa) but not for the bonded case (7.5 MPa). Mann concluded that characteristics of the stem-cement interface had the largest effect on cement mantle stress.

McNamara et al. (1997) discussed the effect of bone-prosthesis bonding on proximal load transfer by using experimental and finite element analysis on a femur. In particular, 3D FE models for an intact femur and a femur implanted with a cementless prosthesis were constructed, and the proximal strains recorded for two loading conditions approximating a one-legged stance: (i) a load applied to the head and an abducting force applied to the greater trochanter; (ii) only a load applied to the head. In modelling the femur, cancellous bone was excluded for simplification. A press-fitted and a fully bonded bone-prosthesis structure were investigated to identify the stem-bone behaviour for both interface conditions and their implications for proximal bone load transfer. Even if the implemented model included some simplifications, authors found excellent correlation between the experimental test strains and predicted strains in FE model for both loading conditions, so they were able to validate their model.

An important field of application of total hip replacement FE models is the preclinical validation of the implant and the study of its primary stability. Viceconti et al. (2001) implemented a complete FE model of bone-implant complex, with the aim to conduct the pre-clinical validation of a new cemented femoral prosthesis – named cement-locked uncemented prosthesis – by using a synergic combination of experimental and numerical methods. The geometry of the femur was created from reconstruction of CT images; materials properties of stem (titanium alloy), cement mantle and femur were modelled as homogenous, isotropic and linear elastic; complete model consisted of 6,347 elements. Two load cases were simulated, a heel strike and a pure torsion. The model was validated against in vitro measurements of bone surface strains as well as against primary stability measurements. Authors focussed on the analysis of micro-movements at cement interface but they also observed that under heel strike loading case the cement was more stressed than under the pure torsion load case. As a main result, the FE analysis predicted, for the new implant device tested, a peak stress in the cement mantle not significantly lower than those reported in other studies on usual fully cemented stems. Even if this study represents only one application of preclinical validation procedure, it gives a clear evidence on the importance of it, in particular when innovative devices must be tested before being implanted.

In the same year and with a similar aim, Stolk et al. (2001) used FE analysis to investigate which muscle groups acting around the hip-joint most affect the load distributions in cemented total hip reconstructions, with a bonded and a debonded femoral stem. The final goal was to determine which muscle groups should be included in pre-clinical tests. The geometry of the femur was obtained from CT scans of the proximal part of a human femur; the cement mantle had a minimal thickness of 2 mm; both bonded and debonded stem were simulated; 19 muscle forces were included in the model and a hip joint force was applied to the centre of the prosthetic head. They considered force magnitudes throughout the gait cycle scaled to a body weight of 735 N. The model consisted of 2,226 brick elements. Results suggest that a loading configuration including the hip-joint contact force and the abductor forces can adequately reproduce in vivo loading of cemented total hip reconstructions in pre-clinical test.

Pancanti et al. (2003) investigated the primary stability of a cementless stem implementing a FE model of a proximal part of human femur with a cementless stem and without considering muscle forces in the model. Muscle forces were not considered because the aim of the simulation was limited to predict the peak micro-movements over the entire bone-implant surface in four patients performing nine different tasks. The model consisted of 9,349 elements and frictional contact was modelled at the bone–implant interface. Among all patients, the largest value of micro-motion, 107 μm, was predicted for a patient performing the stair-climbing task. Interestingly, the distribution of micro-motion seemed not to only depend on the bone-implant geometry but also on how each subject performed the motor task; thus, in the context of pre-clinical validation, the implemented FEA may help to investigate the sensitivity of measurements to the inter-subject variability.

Andreaus et al. (2008), Andreaus and Colloca (2009) conducted a similar study based on a 3D FE model of a physiological and a prosthetic human femur obtained by CT images; a CAD software application was then used to implement an intervention of virtual surgery. The physiological model consisted of 25,637 elements, while the prosthetic model was made of 77,219 elements. Loading was simplified by grouping the action of functionally similar hip muscles. An instant of maximum muscular activity and high joint contact force was then selected for the activities of walking and stair climbing. Numerical simulations – conducted with Comsol 3.3 – showed dramatic variations of maximum and minimum value of stress and strain energy density when comparing walking with stair climbing, even though the spatial locations on the model remained unchanged. Stair climbing was confirmed to be a critical task for primary stability of the prosthetic femur.

Complete models of femur with a prosthesis inside are also used to study the fatigue and fracture of the cement mantle of cemented hip implants in more realistic configurations which take into account for more than only one sample of specific material. Grasa et al. (2005) proposed a new model to predict the reliability of fatigue function of the cement by using a stochastic methodology which allowed to analyse the damage accumulation. The FE model consisted of only proximal part of the femur, made of 11,647 hexahedral bilinear elements; cement-bone interface and cement-stem interface were completely bonded. As for loading conditions, forces developed during walking and stair climbing were taken into account, also including forces acting on the prosthetic head and muscle forces. Analyses were performed with Abaqus v.6.3. After 25 million of loading cycles, a higher failure probability was observed for the specific examined stem with respect to some limited areas located distally in the model; the result was mainly due to the distal cement stress concentration. This interesting methodological study did confirm – as other works had only predicted – that stair climbing is far more detrimental than walking.

Jeffers et al. (2007) conducted a similar study to investigate the ability of a computational method to predict fatigue cracking in experimental models of a simplified implanted femur structure. Computational models were generated of the same geometry as the experimental specimens (CAD reconstruction), with residual stress and porosity simulated in the cement mantle; steel was chosen for stem and aluminium for the external fixture, the latter having a geometry similar to the proximal part of the human femur. Loading was applied to represent level gait for two million cycles. Cement fatigue and creep were simulated using the continuum damage mechanics. Complete cement mantle fractures occurred in the FE models and in the experimental specimens in a similar location. Results of the study suggested that the continuum method is able to predict fatigue locations in a timescale which is comparable with experimental duration; this is an interesting result, notwithstanding simplifications and limitations of the used model.

The scientific peer-reviewed literature also included some studies on the accuracy of total hip replacement surgery (THR) and on the effects of implant positioning on the initial stability. Three of them are briefly described and commented here below.

Lattanzi et al. (2003) investigated the repeatability of orthopaedic surgeons' decisions in planning THR surgery, and compared planned accuracy *vs* accuracy achieved by implementing a conventional unassisted surgical procedure. Three surgeons planned the surgery through a CT-based preoperative planning software. The difference between the implant position achieved at the surgery time and the previously planned position was computed. Results showed that, in most cases, the surgeon positioned the stem more distally and more posteriorly in the pre-operative planning with respect to the stem position during the successive surgery.

In the Reggiani et al.'s study (2008), two FE models of an implanted femur and a titanium alloy stem implant were created, one with the stem positioned according to the pre-surgery planning and the other with the stem fixed in the surgically achieved position (the model had been validated in a previous work by the same author, Reggiani et al. (2007), comparing experimental micro-motion measurements). With a torque of 11.4 Nm applied to the proximal part of hip stem, the planned model predicted a peak Von Mises stress in the bone of 10 MPa; a 12 MPa stress was predicted by the achieved model at the same location. Differences in this case were small because differences between planned and achieved position were quite small, too. However, the study confirms the relevance of prediction of primary stability.

Recently, Bah et al. (2011) used a FE model to describe a statistical investigation into the effects of implant positioning on the initial stability of a cementless total hip replacement. Mesh morphing was combined with design of computer experiments to automatically construct finite element meshes for a range of pre-defined femur-implant configurations and to predict implant micro-motions under joint contact and muscle loading. In particular, a stem of titanium alloy was rotated by $\alpha = [-12°,8°]$, $\beta = [-2.4°,3.5°]$ and $\gamma = [-2°,0.8°]$ around the Z, Y and X axis of a local coordinate system, to replicate retro-anteversion, varus–valgus and ante-posterior orientations. The FE models were subjected to joint and abductor loads during normal walking mode, applied at a single node at the centre of the top surface of the head of the implant and on the greater trochanter, respectively. A 4-noded tetrahedral volume mesh was produced using ANSYS11. The percentage of implant area with micromotion greater than the limit value of 50 μm were considered for THR performance assessment. As a result, the highest percentage was found when the implant was rotated towards the anterior part of the femur, i.e. around the medial–lateral axis; implant maximum micromotion was more dependent on antero-posterior (45%) and varus–valgus (43%) than ante-retroversion rotation. The best orientation angles were found to be: $-11.83°$ of retroversion,

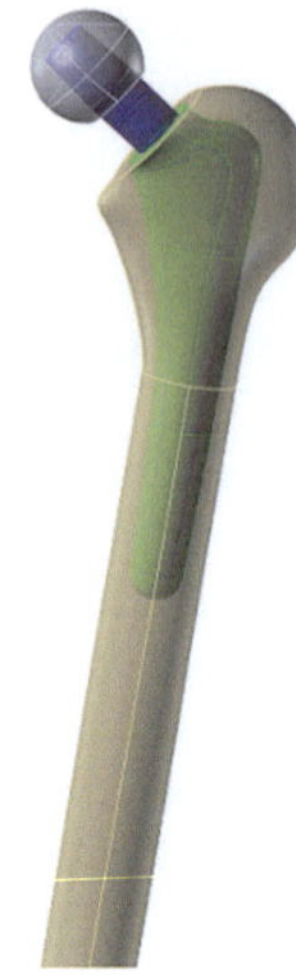

Fig. 3 Complete model created to implement finite element analysis: Model2

2.98° of valgus and 1.67° of posterior orientation. Even though the range of orientation angles was quite limited, and configurations were only compared in terms of micromotions, this study gives a good understanding about the effects of different implant positioning.

Based on the above studies, and with the aim to evaluate the effects of the stem positioning in a configuration close to real situation (Viceconti et al. 2001), in the present study we designed a second Model, named "Model2", consisted of: the same femoral stem used for Model1 – i.e. the Model comparable with the experimental test configuration – , fully bonded in a 3 mm cement mantle as in the literature (Lee et al. 1993; Mann et al. 1995; Ramaniraka et al. 2000) and the external femur reconstructed dimensionally according to Lattanzi et al. (2003) as shown in Fig. 3.

The proximal part of the femur was modeled as in the relevant literature (Mann et al. 1995; McNamara et al. 1997; Stolk et al. 2001; Pancanti et al. 2003); in particular, it consists of the proximal part of the femur as in Grasa et al. (2005). For Model2, two configurations have been considered, the former with the stem oriented according to ISO Standard positioning, the latter according to the critical positioning previously individuated. Boundary and loading conditions were the same as in the previous Model1 – loading applied only at the centre of head prosthesis, interfaces stem-cement and stem-bone completely bounded – . Mesh characteristics of Model2 consisted of 28,150 tetrahedral elements and 123,531 degrees of freedom as shown in detail in Table 2. The mechanical properties of the materials used in the simulation are the same as for Model1, but in Model2 the femur was modeled instead of the metallic fixture; values are presented in Table 1. Material properties of cortical bone were chosen according to Andreaus et al. (2008). Even if, as emerged by literature review, the presence of muscle forces is an important parameter to take into account in order to reproduce in vivo loading (Stolk et al. 2001), the present study was conducted in agreement with Pancanti et al.'s study (2003), thus not including muscular forces in the model in order to simplify it at least at this first stage of the study.

To summarise, two main models have been created to conduct the present study: the first one, called "Model1", to replicate the mechanical test configuration and to simulate standard and critical orientation of the stem; the second one, indicated as "Model2", which includes the femur with prosthesis inside, to simulate more physiological configurations.

3 Results

3.1 *Numerical Simulations*

3.1.1 Model1: Identification of Critical Configuration

Main aim of this part of the study was to discover whether and to which extent a different stem orientation may influence the variation of maximum stress, thus identifying a more critical situation for the prosthesis implant. Therefore a series of finite element analyses was implemented by using the Model1 with the stem positioned under different orientations. In the first set of simulations, as explained in the Methods Section, the angle α was varied from 2° to 15° while angle β was kept fixed at 9° as indicated in the above ISO Standard. In the second set of simulations, angle α was kept fixed at 10° while the angle β was varied from 2° to 13°. The above ranges were selected on the basis of clinical indications of possible implantation angles and taking into account geometrical limitations due to the specific constraint and the metallic fixture associated with the model. From the results obtained in the first set of simulations, it is evident that the Von Mises stress on the stem increases with the decrease of angle α. On the other hand, the results associated with the second set of simulations show the increase of the Von Mises stress with the increase of β angle values (Fig. 4).

From the analysis of the above results, and assuming that the effects of angle variation along the two anatomical planes may be linearly combined, a particular critical positioning has been identified for the stem: more specifically, in each set of simulations the alpha and beta angle was selected as the value in correspondence of which the maximum value of Von Mises stress had been obtained. Thus, the critical orientation of the stem has been identified by $\alpha = 3°$ and $\beta = 12°$. Simulations have been then conducted for both the standard ($\alpha = 10°$ and $\beta = 9°$) and the critical configuration. Table 3 reports the values of stress – Von Mises stress, first principal stress and normal stress-z direction – and displacement obtained for the two configurations by applying the same load (3 kN) and by using the same D constraint distance (52 mm, according to ISO 7206-4: 2002).

It is worth to notice that, under the specified load, Von Mises stress increases of 35.4% and compression z-component of 31.5% for the critical with respect to the standard configuration. Figure 5 shows Von Mises stress distribution on the stem surface in correspondence with the critical configuration.

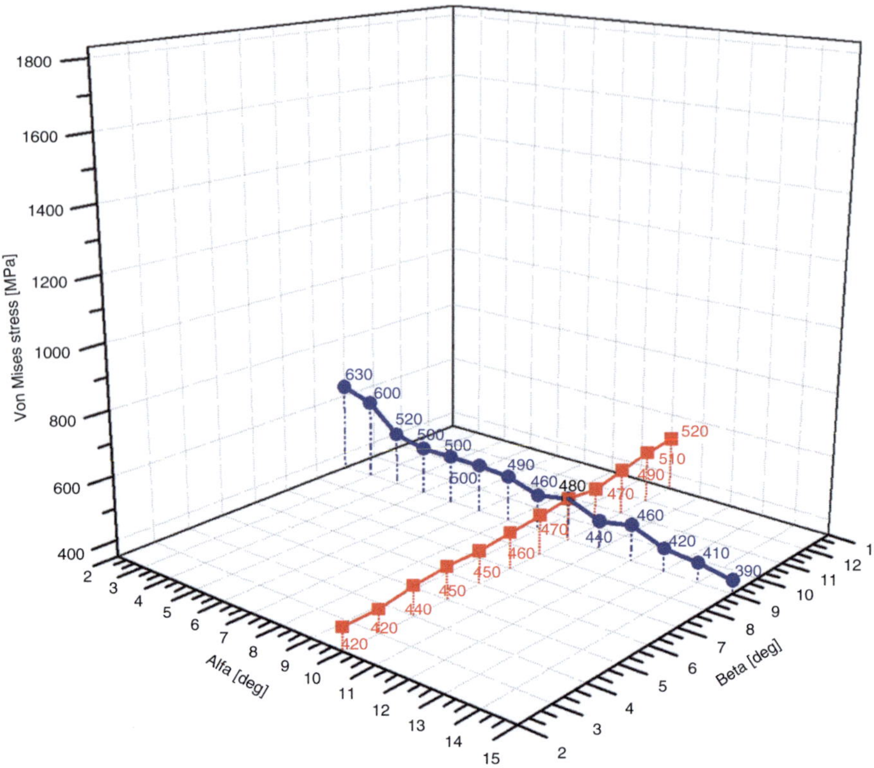

Fig. 4 Model1: maximum Von Mises Stress in correspondence with different angular positions. Level of stem constraint was in compliance with ISO 7206-4: 2002

Table 3 Comparison of maximum stress: standard and critical stem positioning; values refer to the stem subdomain

Model1 (D = 52 mm)					
Load	Configuration	$\sigma_{Mises,\,max}$ [MPa]	$\sigma_{I,\,max}$ [MPa]	$\sigma_{z,\,max}$ [MPa]	Displacement in z direction [mm]
−3,000 N	Critical	650	610	Compression: −710 Traction: 590	−1.27
	Standard	480	430	Compression: −540 Traction: 430	−0.84

A stress concentration area is present in medial-anterior stem portion at the upper surface of the cement. In this area, the normal stress component in z direction, σ_z, is also maximum in compression. Other simulations have been then performed to identify the proper load to apply in mechanical test to create a more critical – even though physiologically reasonable – condition for the stem. In particular, a compression load of 5 kN associated with the critical configuration, delivered a value of Von Mises stress greater than maximum yield stress for the stem

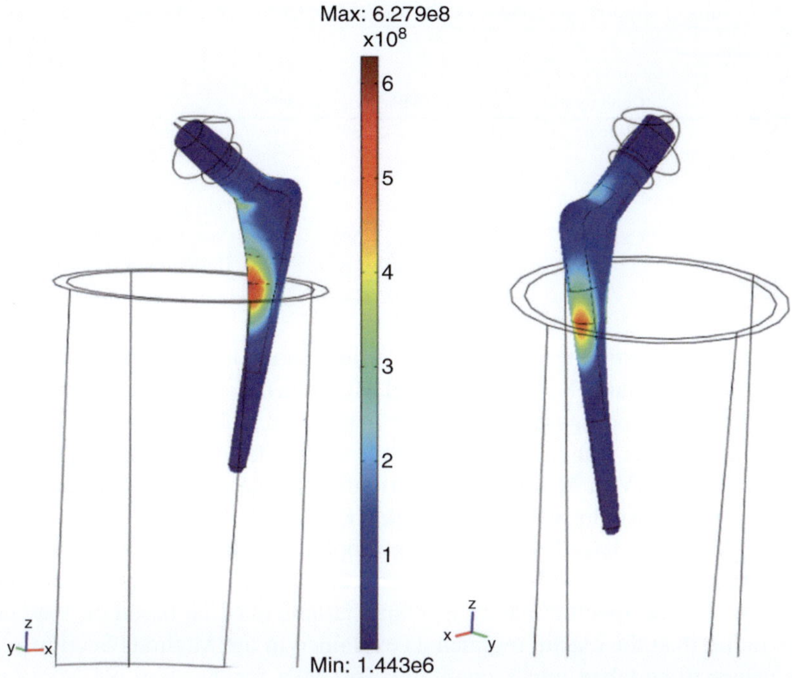

Fig. 5 Model1: Von Mises stress distribution on the stem surface in critical configuration ($\alpha = 3°$, $\beta = 12°$), load 3 kN, ISO 7206-4:2002

material, i.e. 860 MPa (Chao and Lòpez 2007; Costa et al. 2006). In this case, a simulation in elastic-plastic regimen has been conducted besides the standard simulation in elastic regimen.

3.1.2 Model2: Finite Elements Analysis

As explained in the Methods Section, Model2 has been created with the objective of evaluating the effects of an implant configuration closer than that investigated by using Model1 to real situation.

Similarly to Model1 application, also for Model2 two different configurations have been analyzed, with the stem oriented according to the standard positioning – $\alpha = 10°$ and $\beta = 9°$ – and the critical positioning – $\alpha = 3°$ and $\beta = 12°$ -, respectively.

As a first result, we can observe that, even though Model2 cementation is less critical than Model1 because the stem is embedded in the cement mantle up to the collar (osteotomy plan) rather than at a certain level along the stem, again a different stem orientation does determine higher stress values. In particular, as shown in Table 4 for the critical *vs* the standard configuration: (i) Von Mises stress in correspondence with the stem neck increases by 19.8% – from 560 to 671 MPa – ; (ii) z-component in compression increases by 2.26% – from −486 to −497 MPa – ; (iii) the first component of stress increases by 35.6% – from 278 to 377 MPa – . Total

Table 4 Results of simulations conducted for Model2; values refer to the stem subdomain

Model2					
Load	Configuration	$\sigma_{Mises,\ max}$ [MPa]	$\sigma_{I,\ max}$ [MPa]	$\sigma_{z,\ max}$ [MPa]	Displacement total and in z direction [mm]
−3,000 N (elastic regimen)	Critical	671	377	Compression: −497 Traction: 233	Total: 10.90 Z: −3.80
	Standard	560	278	Compression: −486 Traction: 150	Total: 8.79 Z: −2.71
−5,000 N (elastic-plastic regimen)	Critical	860	593	Compression: −727 Traction: 394	Total: 18.90 Z: −6.57

displacement, too, estimated through simulations, shows higher values in the critical configuration (standard: 8.79 mm, critical 10.9 mm). When referred to a clinical context, the above findings may play a role in the stem mobilization, the consequent detachment from the cement mantle and the eventual failure of the implant.

Figure 6 shows Von Mises stress distribution in the critical configuration; the location of the maximum value is highlighted.

The above results have been obtained by applying a maximum compressive load of 3 kN. Once a load of 5 kN is applied, and the stresses reasonably overcome the yield stress for the specific materials, computations must be based on the elastic-plastic rather than the elastic regimen as explained in the Methods Section – yield stress values to be taken into account are: 860 MPa for the stem; 94 MPa for the cement; 114 MPa for the femur cortical bone – . The results of this second analysis are presented in Table 4, too. As expected, it can be observed that the calculated Von Mises stress is higher than the yield stress for all the involved materials. The maximum value of the first principal stress on the stem is 593 MPa and the value of z-component in compression increases up to −727 MPa. Under this condition, total and z-component displacement also show high values, namely 18.90 mm for total displacement and −6.57 mm for z-direction displacement.

3.2 *Fatigue Experimental Test*

3.2.1 Test Main Characteristics, Procedure and Results

As explained in the previous paragraphs, after the performance of the planned set of numerical simulations, a critical stem positioning was identified. The successive step was then carried out by the implementing the mechanical test under the critical situation, so as to: (i) have a further validation of the simulation approach during the first cycles of the mechanical test; (ii) investigate the consequence of the estimated Von Mises stresses on the lifecycle of the stem.

The mechanical fatigue test was thus implemented by using the MTS test system, as described in the Methods Section and according to the ISO 7206-4: 2002 but for the angular position. The stem was in fact positioned in the critical position represented by the adduction angle $\alpha = 3°$ and by the stem flexion angle

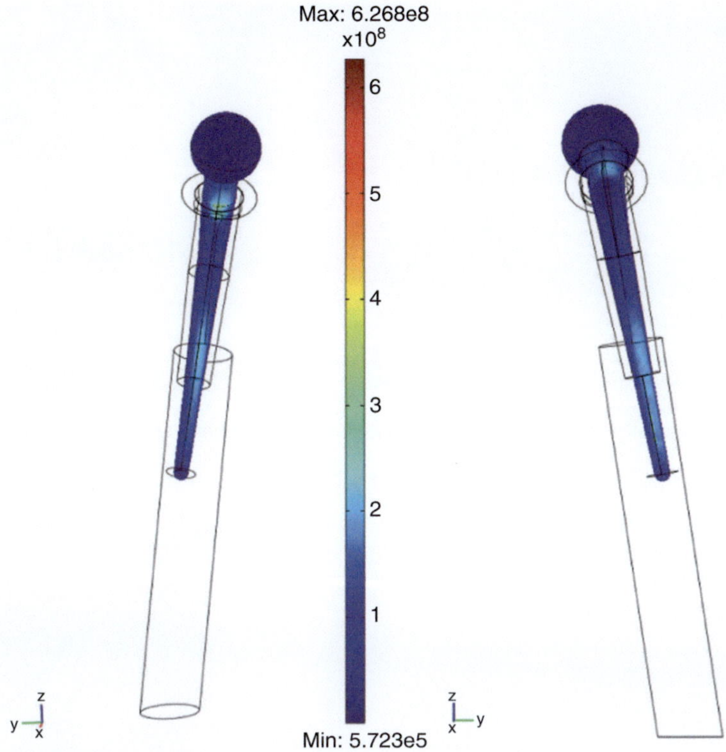

Fig. 6 Model2: Von Mises stress distribution on the stem surface in critical configuration ($\alpha = 3°$, $\beta = 12°$), load 3 kN

$\beta = 12°$. The stem was then accurately cemented, and a fatigue test procedure was uploaded and performed, based on a cyclic load varying from 300 to 5,000 N (compression mode) with a 10 Hz frequency.

The results of the test may be analyzed by splitting the test in two parts: (i) the first part of the test was made of 1,838 cycles, during which the stem did not present visible alterations but a primary fracture of the cement mantle was observed at its lateral aspect in contact with the stem; (ii) the second part of the test, made of 1,948 cycles, showed a critical increase of the measured displacement and, consequently, of the deflection of the femoral head, due to the separation occurred at the interface between cement and stem and to the partial mobilization of the stem (Fig. 7). The fatigue test was stopped after a total amount of 3,786 cycles because the head deflection reached the fixed threshold of 5 mm.

3.2.2 Examination of the Test Specimen

At the end of the mechanical test the stem was extracted from the mantle of cement in which it was embedded. A new, deformed configuration was immediately evident, as well as its separation from the lateral aspect of the cement.

Fig. 7 Fracture of cement mantle during the fatigue test under critical configuration (ISO 7206-4: 2002)

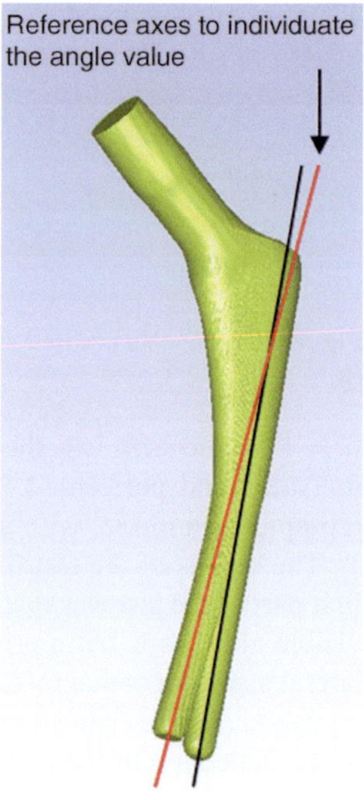

Fig. 8 Comparison between scanned images of the femoral stem before and after the fatigue test

A more accurate analysis was then conducted on the stem surface by using a microscope (zoom 200x) but no superficial abnormalities were observed. The stem was scanned again and by using the software Leios (EGS s.r.l., Italy) it was possible to compare the stem geometry scanned before and after the mechanical test, as shown in Fig. 8. The major difference is evident in the distal part of the stem. More specifically, and with respect to the stem axis, an angular difference of 6.7° was measured between initial and deformed femoral stem.

4 Discussion

4.1 Model1: Numerical Simulation and Fatigue Experimental Test

Comparison between FEM simulation and mechanical test results can be conducted on the basis of one of the most relevant parameters to monitor during a stem fatigue test, namely the deflection of the head. During the performed mechanical test, the deflection was acquired through the MTS displacement actuator, while in the FEM simulation it is delivered as the head displacement in z direction.

In order to properly perform the comparison, FEM simulations were implemented considering that in the first part of the test the stem was correctly constrained in the cement without deformation, thus at D = 52 mm, whilst in the second part it was mobilized and deformed, and the new constraint level should properly indicated as D = 64 mm (measured after the mechanical test):

– first FEM simulation: Model1, critical positioning; distance D = 53 mm (1 mm was added with respect to the theoretical value to properly take into account the small lowering of the level of cement after polymerization); applied load = 5 kN; simulation in elastic regimen. Main results were: z displacement = −2.21 mm; maximum Von Mises stress = 1,200 MPa;
– second FEM simulation: Model1, critical positioning; distance D = 64 mm; applied load = 5 kN; simulation in elastic-plastic regimen (yield stress for the stem = 860 MPa). Main results were: z displacement = −4.46 mm; Von Mises stress = 860 MPa.

For the other stress components (see Table 5): the same value of −1,200 MPa was obtained for the z-component stress in compression in both simulations; the first component of stress increased of 20% from first to second simulation.

As shown in Table 5, the difference between the displacement measured by the MTS System during the mechanical test and the displacement estimated through the FEM simulations is constant: the FEM simulation underestimated the measured displacement value of 0.4 mm. Based on the above results, it is reasonable to consider the FE analysis properly conducted and of a certain relevance in the identification of implant positioning at a risk of mechanical failure.

4.2 Comparison Between Model1 and Model2

The rationale for the creation of Model1 was the aim of reproducing the same configuration used in the mechanical fatigue test according to the ISO Standards; thus, the model has a correct correspondence with the mechanical test chain and, from a clinical point of view, it allows the detection of potentially critical

Table 5 Comparison between results obtained in mechanical test and numerical simulations

Model1 Load Distance D	Configuration	$\sigma_{Mises,\,max}$ [MPa]	$\sigma_{I,\,max}$ [MPa]	$\sigma_{z,\,max}$ [MPa]	Displacement in z direction [mm]
Fatigue test: first part −5,000 N D = 53 mm (elastic regimen)	Critical	1,200	1,000	Compression: −1,200 Traction: 1,000	−2.21
Delta displacement actuator of MTS machine: −2.6 mm					
Fatigue test: second part −5,000 N D = 64 mm (elastic-plastic regimen)	Critical	860	1,200	Compression: −1,200 Traction: 1,200	−4.46
Delta displacement actuator of MTS machine: −4.85 mm					

configuration which, as previously discussed, might become dangerous once a first mobilization has occurred at the interface between stem and cement – or between stem and bone – in a "in-vivo" implant. This possible "domino effect" is often underestimated in clinical practice but represents a clear indication on the need of studies like the present one. Correlated with the analysis of specific prosthesis design, the Model1 FEM simulation can be valuable to early understand whether the stem is working in a stress level near the endurance limit of the material that would inevitably culminate a proximal loosening and a fatigue failure; an example of this application was presented and discussed in the study of Griza et al. (2008a).

The rationale behind the development of Model2 was rather the attempt to evaluate the consequences, in a configuration closer to a clinical setting, of the stem positioning individuated through Model1 FEM simulations. In particular, Model2 represents the usual clinical configuration of the hip prosthesis implant in an ideally functioning environment, i.e. without any other correlated event that may occur and damage the implant during its lifetime. It is here mandatory to specify, however, that Model2 does not attempt to exactly recreate the in vivo mechanical environment, i.e. muscle forces acting of the femur, and different materials properties of the trabecular and of the cortical bone were not taken into account in the FEM simulation. In the present study the above limitations have been considered acceptable, since focus of the study was only put on the effects of different angular positions on the maximum Von Mises stresses experienced by the stem, and in a first approximation the lines of force of the main involved muscles were assumed invariant with respect to the "limited" variations in the stem angular position.

When comparing results obtained by the two Models FEM simulations, it is well evident how much the stress distribution on the stem changes in the two cases both in terms of spatial distribution and of intensity: while Model1 FEM simulations prove that after the loss of the primary stability intense stress values are allocated along the stem in correspondence with the constraint level, Model2 FEM simulations show that the most stressed area of the stem during the first part of the implant active lifetime, when the implant is solid with the bone or the cement interface, is the area at the neck of the stem, at the level of the osteotomy plane. Abnormalities in the angular position of the stem – within the angular variations examined in the present study, of course – as for both hip adduction and flexion do not change the most stressed areas in both Models; however, interesting changes can be observed in the variation of stress values. In fact, even though the maximum Von Mises stress in critical configuration – adduction angle $\alpha = 3°$, flexion angle $\beta = 12°$ – shows very similar values for the two Models – 650 MPa at the constraint level in Model1, 671 MPa at the osteotomy level in Model2 – , relative percentage values are quite different when comparing standard and critical configuration: (i) Model1 increase of Von Mises stress is more than 35% for the critical configuration; (ii) Model2 increase is around 20%. This means that a certain misalignment during surgical implantation might entail some acceleration in the loss of primary stability but, even more important, is responsible for a faster damage of the whole implant once the stability is lost.

4.3 Limitations of the Study and Future Plans

The study presented in this chapter investigates the potentialities of research studies in the field of hip prosthesis implants using FE modeling and experimental fatigue tests in order to analyze potential risks of implant mechanical failure. To this aim, attention has been paid to the relevance of a correct positioning of the stem and to the identification of potential critical configurations which may cause or play a major role in the implant failure due to fatigue failure. Results hereby presented, however, are affected by some limitations and bias which are briefly reported here below.

As a first limitation, the Model1 critical configuration has been identified by alternatively changing values of α and β angles – i.e. one was kept fixed at the ISO Standard value while the other was varied – with a step of 1° in a defined angular range (2–15° for α, 2–13° for β). Some angular values were excluded because it was impossible to implement them for geometrical problems correlated with the mechanical model, i.e. it would be impossible to cement the stem in the metallic fixture; some others were also excluded because unlikely to be reproduced in a clinical environment. Moreover, the identification of the most critical positioning of the stem was obtained by selecting, in both groups of simulations, the angle value in correspondence of which the maximum Von Mises stress has been obtained. This means that not all the possible combinations of α and β in the selected angular ranges have been implemented, and that a linear effect superimposition has been hypothesized in a first approximation. In a future development of the present study, all feasible configurations will be analyzed and their critical aspects investigated, so as to validate or rebut the above hypothesis of linear behavior.

As for Model2, as previously indicated, other limitations must be highlighted i.e. boundary conditions and material properties of the bone – only standard cortical bone properties have been included in the Model – . Such limitations have been considered acceptable in the present study being its main aim the observation of the macro effects of differences in Model1 positioning in a situation similar to the early phases of the clinical implant represented by Model2. Interesting future developments of Model2 will be soon taken into consideration; as an example, in the present study, only the situation has been analyzed of a cemented stem including a 3 mm cement mantle between the stem and the bone, while it will be interesting to modify the Model by excluding the cement so as to study whether the effect of different angular positioning is more or less relevant in implants without cement, not to mention the interest in the effects induced on the bone.

4.4 The 2010 Revision of the International Standard ISO 7206-4

Simulations and the mechanical test discussed in the present study for Model1 were based on the ISO standard 7206-4 version published in 2002 (ISO 7206-4: 2002). However, in 2010 a revision of Part 4 of the Standard has been published by the International Standard Organization (ISO 7206-4: 2010). Within the purpose of the present study, the main difference between the two versions deals with the value of distance D – namely the distance between the center of the head and the cement level – at which the stem is constrained: in the 2002 version of the Standard, the value of D was parameterized according to the stem dimension, whilst in the 2010 version the stem we have analyzed must be embedded at a distance $D = 80$ mm, as already indicated in the previous 1989 version of ISO 7206-4 (ISO 7206-4: 1989). This embedding requirement does represent a more critical test condition for small stems, as is the case for the stem under examination in the present study. Thus, it is reasonable to hypothesize that, according to the 2010 version of the Standard, most of the findings of the present study overestimate the lifetime of an implanted but mobilized stem. To have a better idea of the expected results, some preliminary numerical simulations have been conducted with the same Model1 of the stem as in the present study; the same method has been also applied to alternatively vary the angle α from 2° to 15° with $\beta = 9°$ fixed, and successively to vary the angle β from 2° to 13° with $\alpha = 10°$. At the end of this set of simulations a critical condition for the stem was identified in correspondence with $\alpha = 3°$ and $\beta = 12°$, based again on the evaluation of Von Mises stress maximum values and by excluding some positions like $\alpha = 3°$ with $\beta = 13°$ which could not be implemented for geometrical limitations. Figure 9 summarizes the above results.

Interestingly, the critical configuration is the same previously obtained in compliance with the 2002 version of the ISO Standard 7206-4. However, in this last case, and for a standard load of 3 kN, Von Mises stress increases of 48.5% with respect to the standard configuration, from 1,173 to 1,742 MPa, and z-displacement component in compression increases of 62.2%, from −2.89 to −4.69 mm. It is worth to observe that by applying the 2010 version of the Standard, not only the percentages are higher than in the previous analysis, but also absolute Von Mises stresses and z-displacements are already very high under the standard configuration: for standard positioning, in fact, Von Mises stress increases of about 144% and

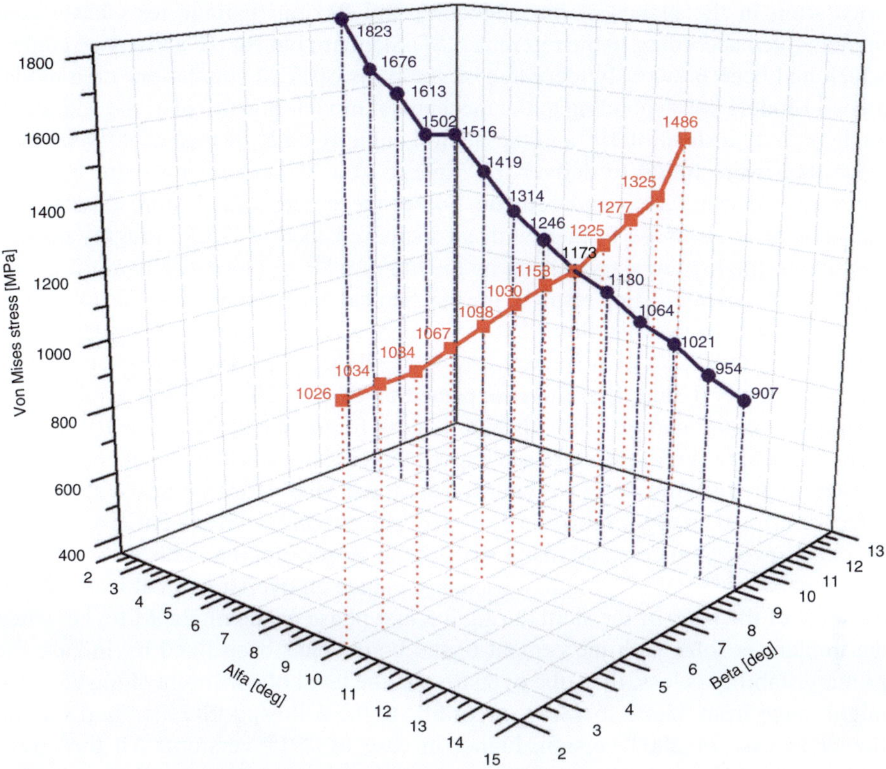

Fig. 9 Model1: maximum Von Mises Stress in correspondence with different angular positions. Level of stem constraint was in compliance with ISO 7206-4: 2010

z-displacement of about 244% when comparing the 2010 with the 2002 version of the Standard. On the basis of the critical results of the FEM simulations, and keeping in mind that the numerical simulations resulted slightly conservative with respect to the experimental tests when the 2002 version had been implemented, it is expected that the stem has a significantly shorter resistance to fatigue and clinical failure. Mechanical tests are almost ready to verify how close numerical simulations are to the mechanical behavior of the femoral stem. Eventual consequences from the clinical point of view will be investigated in depth.

5 Conclusions

The aim of the present research was to study mechanical properties and lifetime of hip prosthetic implants by using two complementary methods, i.e. mechanical fatigue tests with a test-and-measure device and numerical simulations based on FE analysis, to investigate potential critical implant configuration in terms of orientation in the planes of hip adduction and flexion. Fatigue tests have been implemented according to the reference ISO Standard but for the above two angles, which had been previously identified on the basis of FEM simulations on a model named Model1 corresponding to the mechanical model. In particular, the adduction angle $\alpha = 3°$ instead of 10° and the flexion angle $\beta = 12°$ instead of 9° were used since they were found to deliver the highest Von Mises stresses at the level of constraint between stem and cement. To set the configuration more critical, the mechanical test was performed with an increased load of 5 kN, roughly corresponding to the loading experienced by the hip of a 80 kg human being during stair climbing. As an interesting result of the mechanical experimental test, the examined stem underwent a mobilization from the cement after less than 2,000 cycles and an irreversible deflection of 6.7° after less than 4,000 cycles (according to the ISO Standard, femoral stem lifetime is expected to be 5×10^6 cycles at a minimum, with a sinusoidal load of 2,300 kN at a loading frequency in the range 10–30 Hz). Agreement between FE Model1 and experimental test was good enough to encourage the implementation of a second Model named Model2 to better investigate the effect of critical configuration in a more realistic clinical environment. This second Model allowed to understand that a wrong angular positioning, with angular variation in between 3° and 7°, might reasonably induce an overstress of at least 20% at the level of the neck of the stem during the first phase of the implant life, i.e. when the implant is solid with the cement or the bone it has been fixed to. In case the primary stability is loss, then, the overstress at the level of constraint along the stem might range from 35.4% to 48.5% according to the followed ISO Standard version (lower in case of 2002 version, higher in case of 2010 version). All the above findings should be seriously taken into account from a clinical point of view. In fact, overstress due to overload of the hip prosthesis is not unbelievable, especially when considering the growing increase in obese patients or with body weight greater than a normal-weight patient (Chao 2008; Baleani et al. 1999). Moreover, variation of few degrees in the angular positioning of the femoral stem might be reasonable too, also taking into account abnormalities in the bone architecture of the patients and possible intraoperative bias.

To conclude, the present study proved that the FEM approach might represent a valuable tool of analysis and simulation to support of fatigue experimental tests commonly performed in laboratories, which are destructive and entail associated high costs. For the specific examined stem, the study also showed that the test indications used in the EU normative do represent a proper test configuration, but they are not representative of a possible highly critical condition for the femoral stem. In particular, the study suggested that a possible different orientation of the stem, along with the application of loads greater than those currently set by the Standards, may determine very intense values of stress on the stem that can cause a loss of primary stability and the consequent failure of the entire implant. These considerations, as discussed in different ways in some others works, become relevant in the context of a preclinical validation as well as describing possible clinical scenarios.

Acknowledgments The Authors greatly thank Giorgio De Angelis for his technical support during the experimental phase.

References

Age.na.s HTA Report (2008) Prostheses for primary total hip replacement in Italy. Rome. http://www.salute.gov.it/imgs/C_17_pagineAree_1202_listaFile_itemName_1_file.pdf. Accessed 11 July 2011

Andreaus U, Colloca M (2009) Prediction of micromotion initiation of an implanted femur under physiological loads and constraints using the finite element method. Proc IMechE, Part H, J Eng Med 223:589–605

Andreaus U, Colloca M, Toscano A (2008) Mechanical behaviour of a prosthesized human femur: a comparative analysis between walking and stair climbing by using the finite element method. Biophys Bioengin Lett 1:1–15

Bah MT, Prasanth BN, Taylor M, Browne M (2011) Efficient computational method for assessing the effects of implant positioning in cementless total hip replacements. J Biomech 44:1417–1422

Baleani M, Cristofolini L, Viceconti M (1999) Endurance testing of hip prostheses: a comparison between the load fixed in ISO 7206 standard and the physiological loads. Clin Biomech 14:339–345

Chao J (2008) Is 7206 ISO standard enough to prove the endurance of femoral components of hip prostheses? Eng Fail Anal 15:83–89

Chao J, Lòpez V (2007) Failure analysis of a Ti6Al4V cementless HIP prosthesis. Eng Fail Anal 14:822–830

Costa MYP, Voorwald HJC, Pigatin WL, Guimaraes VA, Cioffi MOH (2006) Evaluation of shoet peening on the fatigue strength of anodized Ti-6Al-4V alloy. Mat Res 9:107–109

Giacomozzi C, Campioni I, De Angelis G, Notarangelo G (2010) Mechanical testing of hip prostheses implemented at the ON TESA testing lab of the Istituto Superiore di Sanità: tools, methods and procedures compliant with international standards. Rapporti ISTISAN 10/36. Is-tituto Superiore di Sanità, Roma

Grasa J, Pérez MA, Bea JA, Garcìa-Aznar JM, Doblaré M (2005) A probabilistic damage model for acrylic cements. Application to the life prediction of cemented hip implants. Int J Fatigue 27:891–904

Griza S, Kwietniewski C, Tarnowski GA, Bertoni F, Reboh Y, Strohaecker TR, Baumvol IJR (2008a) Fatigue failure analysis of a specific total hip prosthesis stem design. Int J Fatigue 30:1325–1332

Griza S, Reis M, Reboh Y, Reguly A, Strohaecker TR (2008b) Failure analysis of uncemented total hip stem due to microstructure and neck stress riser. Eng Fail Anal 15:981–988

Griza S, Zanon G, Silva EP, Bertoni F, Reguly A, Strohaecker TR (2009) Design aspects involved in a cemented THA stem failure case. Eng Fail Anal 16:512–520

Herberts P, Malchau H (2000) Long-term registration has improved the quality of hip replacement: a review of the Swedish THR Register comparing 160000 cases. Acta Orthop Scand 71:111–121

Hodgkinson J, Skinner J, Kay P (2011) Large diameter metal on metal bearing total hip replacements. Official Joined Communication of the British Hip Society and the British Orthopaedic Association. http://www.britishhipsociety.com/pdfs/BHS_MOM_THR.pdf. Accessed 11 July 2011

ISO 5832-3 (1996) Implants for surgery -metallic materials- Part 3: Wrought titanium 6-aluminium 4-vanadium alloy. International Organization for Standardization, Geneva

ISO 5832-4 (1996) Implants for surgery -metallic materials- Part 4: Cobalt-chromium-molybdenum casting alloy. International Organization for Standardization, Geneva
ISO 7206-4 (1989) Implants for surgery – partial and total hip prostheses – Part 4: Determination of endurance properties of stemmed femoral components with application of torsion, 1st edn. International Organization for Standardization, Geneva
ISO 7206-4 (2002) Implants for surgery – partial and total hip prostheses – Part 4: Determination of endurance properties of stemmed femoral components, 2nd edn. International Organization for Standardization, Geneva
ISO 7206-4 (2010) Implants for surgery – partial and total hip prostheses – Part 4: Determination of endurance properties and performance of stemmed femoral components, 3rd edn. International Organization for Standardization, Geneva
ISO 7206-8 (1995) Implants for surgery – partial and total hip prostheses – Part 8: Endurance performance of stemmed femoral components with application of torsion, 1st edn. International Organization for Standardization, Geneva
Jeffers JRT, Browne M, Lennon AB, Prendergast PJ, Taylor M (2007) Cement mantle fatigue failure in total hip replacement: experimental and computational testing. J Biomech 40:1525–1533
Kiefer H (2007) Current trends in total hip arthroplasty in Europe and experiences with the bicontact hip system. In: Sofue M, Endo N (eds) Treatment of osteoarthritic change in the hip, Part IV. Springer Japan, Toyko, pp 205–210
Lattanzi R, Grazi E, Testi D, Viceconti M, Cappello A, Toni A (2003) Accuracy and repeatability of cementless total hip replacement surgery in patients with deformed anatomies. Inform Health Soc Care 28:59–71
Lee IY, Skinner HB, Keyak JH (1993) Effects of variation of cement thickness on bone and cement stress at the tip of a femoral implant. Iowa Orthop J 13:155–159
Lie SA, Havelin LI, Furnes ON, Engesæter LB, Vollset SE (2004) Failure rates for 4762 revision total hip arthroplasties in the Norwegian Arthroplasty Register. J Bone Joint Surg 86-B:504–509
Mann KA, Bartel DL, Wright TM, Burstein AH (1995) Coulomb frictional interfaces in modelling cemented total hip replacements: a more realistic model. J Biomech 28:1067–1078
McNamara BP, Cristofolini L, Toni A, Taylor D (1997) Relationship between bone-prosthesis bonding and load transfer in total hip reconstruction. J Biomech 30:621–630
Pancanti A, Bernakiewicz M, Viceconti M (2003) The primary stability of a cementless stem varies between subjects as much as between activities. J Biomech 36:777–785
Paul JP (1997) Development of standards for orthopaedic implants. Proc Inst Mech Eng Part H 211:119–126
Ploeg HL, Burgi M, Wyss UP (2009) Hip stem fatigue test prediction. Int J Fatigue 31:894–905
Raimondi MT, Pietrabissa R (1999) Modelling evaluation of the testing condition influence on the maximum stress induced in a hip prosthesis during ISO 7206 fatigue testing. Med Eng Phys 21:353–359
Ramaniraka NA, Rakotomanana LR, Leyvraz PF (2000) The fixation of the cemented femoral component. Effects of stem stiffness, cement tickness and roughness of the cement-bone surface. J Bone Joint Surg 82-B:297–303
Reggiani B, Cristofolini L, Varini E, Viceconti M (2007) Predicting the subject-specific primary stability of cementless implants during pre-operative planning: preliminary validation of subject-specific finite-element models. J Biomech 40:2552–2558
Reggiani B, Cristofolini L, Taddei F, Viceconti M (2008) Sensitivity of the primary stability of a cementless hip stem to its position and orientation. Artif Organs 32:555–560
Stolk J, Verdonschot N, Huiskes R (2001) Hip-joint and abductor-muscle forces adequately represent in vivo loading of a cemented total hip reconstruction. J Biomech 34:917–926
Viceconti M, Cristofolini L, Baleani M, Toni A (2001) Preclinical validation of a new partially cemented femoral prosthesis by synergetic use of numerical and experimental methods. J Biomech 34:723–731
Viceconti M, Affatato S, Baleani M, Bordini B, Cristofolini L, Taddei F (2009) Pre-clinical validation of joint prostheses: a systematic approach. J Mech Behav Biomed Mater 2:120–127

Patient-Specific Modelling in Orthopedics: From Image to Surgery

G.T. Gomes, S. Van Cauter, M. De Beule, L. Vigneron, C. Pattyn, and E.A. Audenaert

Abstract In orthopedic surgery, to decide upon intervention and how it can be optimized, surgeons usually rely on subjective analysis of medical images of the patient, obtained from computed tomography, magnetic resonance imaging, ultrasound or other techniques. Recent advancements in computational performance, image analysis and in silico modeling techniques have started to revolutionize clinical practice through the development of quantitative tools, including patient specific models aiming at improving clinical diagnosis and surgical treatment. Anatomical and surgical landmarks as well as features extraction can be automated allowing for the creation of general or patient specific models based on statistical shape models. Preoperative virtual planning and rapid prototyping tools allow the implementation of customized surgical solutions in real clinical environments. In the present chapter we discuss the applications of some of these techniques in orthopedics and present new computer-aided tools that can take us from image analysis to customized surgical treatment.

Keywords Musculoskeletal modelling • Patient-specific models • Surgical planning

G.T. Gomes • C. Pattyn • E.A. Audenaert (✉)
Ghent University Hospital, Ghent, Belgium
e-mail: Emmanuel.Audenaert@UGent.be

S. Van Cauter • M. De Beule
IBiTech–bioMMeda, Ghent University, Ghent, Belgium

L. Vigneron
Orthopedic Department, Materialise NV, Leuven, Belgium

D. Iacoviello and U. Andreaus (eds.), *Biomedical Imaging and Computational Modeling in Biomechanics*, Lecture Notes in Computational Vision and Biomechanics 4,
DOI 10.1007/978-94-007-4270-3_6, © Springer Science+Business Media Dordrecht 2013

1 Virtual Anatomical Landmark Extraction

1.1 Anatomical Landmarks in Orthopedics

The identification of reference parameters or anatomical landmarks is a well-established technique in orthopedic surgery. Anatomical features are used for various applications. Many morphological parameters (e.g. distances, angles) are quantified based on landmarks (Paley 2002). These measurements can serve as a guideline for distinguishing dysplastic from normal morphologies (Delaunay et al. 1997). Also, many studies have shown that accurate prosthetic component positioning is a key factor for the success of joint replacement surgery and have presented recommendations for the orientation angles (Yoon et al. 2008). Joint kinematics is often described by the relative motion of joint coordinate systems that are attached to the bones. These joint coordinate systems can be defined based on anatomical features (Grood and Suntay 1983). Moreover, surgical navigation systems rely on landmarks. Image-based navigation requires patient-to-image registration and this process often relies on registration points that are determined on the image and have to be recognized during the operation (Nizard 2002). Image-free navigation systems use landmarks to create anatomical reference frames that relate the position and orientation of the reference frames that are attached to the patient's bones, to the underlying bony anatomy (Siston et al. 2007). Finally, different landmarks can be used to determine the insertion locations in ligament reconstruction (Schöttle et al. 2007; Ziegler et al. 2010). Anatomical features have thus proven to be applicable throughout all steps of the patient treatment process: diagnosis, preoperative planning, intraoperative navigation and postoperative follow-up.

Anatomical landmarks can be quantified in different ways. On live subjects, they can be located using manual palpation and digitized using a probe. In addition, surgical navigation systems allow to determine landmarks by means of a kinematic analysis of the patient's joint (e.g. centre of the hip, knee and ankle). Software programs are available for manual identification of landmarks on digital medical images and three-dimensional (3D) computer models. Moreover, various techniques for automatic landmark extraction are being developed. An important factor in landmark-based applications is the use of standardized definitions to allow for better result comparison and data exchange (Van Sint Jan and Della Croce 2005; Van Sint Jan 2007).

1.2 Virtual Landmark Identification

Several imaging methods allow users to virtually quantify landmarks on the inner structures of the body. 2D radiography is the most common and widely available imaging technique in orthopedics. However, radiographic images are characterised by the superimposition of anatomical structures and by magnification and distortion errors. Computed tomography (CT) and magnetic resonance (MR) imaging have been used to obtain more detailed, cross-sectional images. Furthermore, multiplanar reformatted images and 3D surface renderings can be generated from these 2D slices.

Most orthopedic systems that use digital images or models require the physician to manually mark the landmarks, relying on his good judgement and experience. Therefore, intra- and interobserver variability is an important factor that should be taken into account when using landmark-based clinical parameters. Several studies have been performed to assess the reproducibility of landmark identification on CT or MR (-based) images and the corresponding morphological parameters (Cerveri et al. 2010; Lerner et al. 2003; Nofrini et al. 2004; Subburaj et al. 2009; Victor et al. 2009; Yoshino et al. 2001), see Table 1.

A direct comparison between studies is not feasible, because of the different methods that are used, but their results indicate that mean variabilities of 2–3 mm/° are not uncommon for some of the landmarks and parameters.

In addition, models have been developed to determine the deviation of clinical parameters that arises from landmark variability. Morton et al. 2007 found a significant variability of absolute kinematic parameters of up to 6.5° and 4.4 mm for tibiofemoral and 7.6° and 6.5 mm for patellofemoral kinematics, assuming standard deviations of up to 2 mm for the anatomical landmark locations. The simulation results of Wolf et al. 2005 indicate that version and abduction errors in acetabular cup alignment can be as large as $\pm$ 10 and $\pm$ 6°, respectively, assuming a 4 mm error in anterior pelvic plane landmark localization. Although these studies used rather high variability for all landmarks involved, they illustrate the significance of estimating the resulting errors in the clinical parameters to preclude misinterpretation of data.

1.3 Automatic Landmark Extraction

Manual localisation of landmarks is inherently characterized by inter- and intraobserver variability. Although computer methods have been developed to assist the user in landmark identification (Hankemeier et al. 2006), the procedure remains time-consuming and a high level of expertise and alertness is required to obtain precise measurements. To overcome these drawbacks, automated landmark extraction techniques are increasingly being used (Ehrhardt et al. 2004; Subburaj et al. 2010). Their validity is usually tested by comparing the parameters obtained by the automated methods to a reference parameter set obtained by manual expert analysis.

We have developed a heuristically driven approach for fast automated landmarks, regions and other parameters extraction (Gomes 2011). Applied to the human femur, it extracts the following parameters: epycondyles (most lateral and medial), transepicondylar axis, condyles (most inferior and posterior locations), intercondylar fossa, femoral condylar axis, centre and radius of the femoral head,

Table 1 Variability of manual landmark identification

Reference	# datasets	# parameters	Images	Variability definition	Averaged over	Observer	Value
Victor et al. (2009)	6	27	MO	DFM	AN and DS	Intra	0.4–1.4 mm
							0.05–1.44°
						Inter	0.3–3.5 mm
							0.08–3.16°
Cerveri et al. (2010)	20	7	MR	DFM	AN	Intra	$\leq$3.0 ± 2.1 mm
							$\leq$3.5 ± 2.2°
						Inter	$\leq$3.1 ± 2.0 mm
							$\leq$3.7 ± 2.3°
Subburaj et al. (2009)	3	14	MO	SD	CA, AN and DS	Inter	2.15–5.98 mm
Yoshino et al. (2001)	33/96	2	SL (axial)	PD	DS	Inter	0.4 ± 0.4° to 1.1 ± 0.6°
Nofrini et al. (2004)	4	4	SV (frontal and lateral) and SL (axial)	PD	AN and DS	Intra	2.8 ± 1.8 to 5.4 ± 2.7 mm
						Inter	3.1 ± 1.6 to 5.4 ± 2.7 mm
Lerner et al. (2003)	7	9	SL (axial and sagittal)	SEM	CA, AN DS and PA	Intra	0.41 mm
						Inter	0.51 mm

Images: *SV* scout view, *SL* slices, *MR* multiplanar reformatted, *MO* 3D model. Variability definition using values from several analyses: *DFM* difference from the mean value, *SD* standard deviation, *PD* pairwise difference, *SEM* standard error of the measurement. Averaged over: *CA* coordinate axes, *AN* analyses, *DS* datasets, *PA* parameters

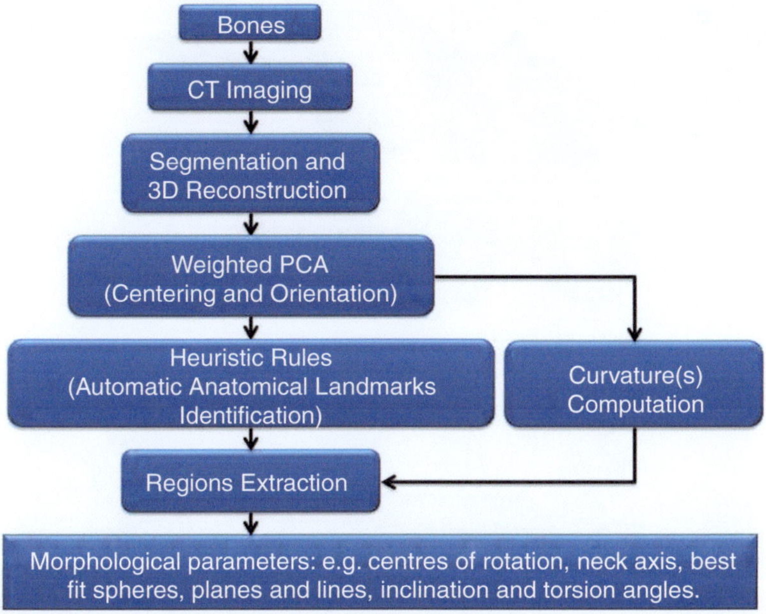

Fig. 1 Overview of the methodology for landmarks and other features extraction

greater trochanter (most lateral and most superior locations), lesser trochanter, quadrate tubercle, iliofemoral ligament insertions, trochanteric fossa, linea aspera, abductor tubercle, anterior and posterior cruciate ligament insertions, axis of the neck, isthmus of the neck, optimum antegrade and retrograde nail insertions from least squares B-spline fit to the shaft medial points or from least squares elliptic hyperboloid fitting to the shaft and Bezier curve computation from the hyperboloid's medial axes (Gomes 2011; Van Cauter et al. 2012), inclination and version of the head and bending between distal and proximal parts of the shaft (Figs. 1 and 2). A high resolution bone reconstructed from sub millimetre CT voxels (with over 50,000 vertices), can be fully analysed in less than 10 s. on an average desktop machine running Matlab. The algorithms have been tested using a database of 100 CT scanned femurs with wide normal and some pathological variability.

These automated techniques offer several advantages over manual analyses. They may save time for the physician and allow rapid data processing to obtain morphological measurements for large control and patient populations. Moreover, they eliminate observer variability by using standardized procedures, which may result in an improved (e.g. pre- and postoperative) data comparison and exchange. Their validity and robustness should always be tested using large datasets. It should also be taken into account that standardized algorithms may not always be able to correctly capture the features in severely deformed anatomies but this can be addressed with the use of shape models that capture such variability.

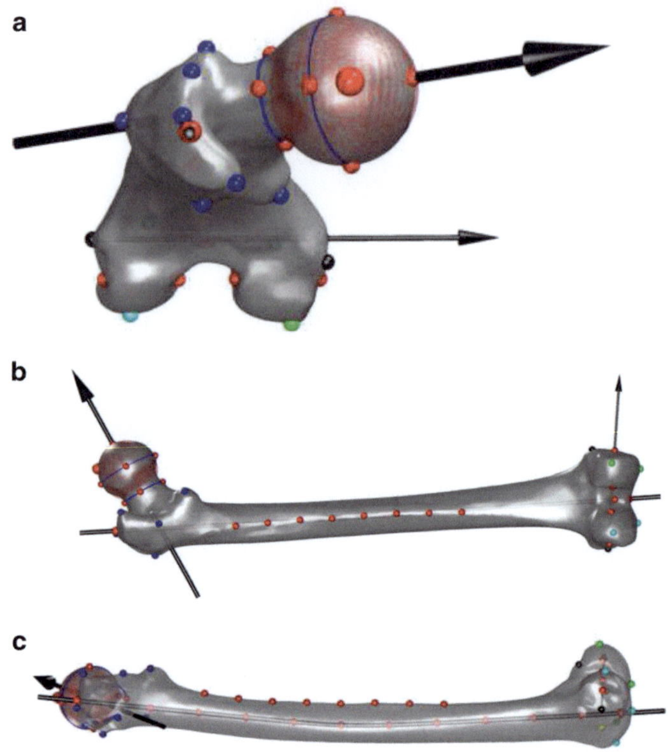

Fig. 2 Superior (**a**), posterior (**b**) and lateral (**c**) views of a *left side* femur with extracted features i. e. landmarks, axes, best fits (Gomes 2011)

2 Statistical Shape Models

In medical imaging, statistical shape analysis provides an important and increasingly popular means for computing 3D models of the patient's anatomy. This capability has been used for:

- 3D reconstruction of anatomic structures from sparse patient-specific input data (Barratt et al. 2008; Benameur et al. 2003; Blanz et al. 2004; Fleute and Lavallée 1998; Kurazume et al. 2009; Rajamani et al. 2007; Stindel et al. 2002; Zheng et al. 2006; Zheng et al. 2008; Zheng and Schumann 2009). While 3D image datasets (e.g. CT, MR) can be acquired for segmentation purposes, the constraints of clinical reality (e.g. time, cost and exposure to unacceptable levels of ionizing radiation) preclude their widespread use, on the other hand 2D medical imaging modalities such as X-rays, ultrasound or fluoroscopy, or direct pointer digitization can be used to derive patient-specific information pre- or intraoperatively from the patient's anatomy for the purpose of 3D reconstruction. The acquired patient-specific data can be processed and used to deform a shape model to finally reconstruct the patient's anatomy

- Predicting the shape of one bone from the observation of another from the same joint (Yang et al. 2008)
- Segmentation (Baldwin et al. 2010; Behiels et al. 2002; Heimann and Meinzer 2009; Tang and Ellis 2005), SSM allows encapsulation of prior anatomical knowledge for compensating low contrast and/or high levels of noise in the images, such models can achieve robust segmentation by constraining the possible shapes (Cootes and Taylor 2004; Cootes et al. 1995)
- Design of prosthesis and biomechanical finite element analysis (Bryan et al. 2009)
- Aiding in the detection of pathologies related to shape (e.g. cam impingement, osteoarthritis), anatomical differences related to sex and aging (Styner et al. 2005).

2.1 Basic Concepts

Statistical shape models (SSM) aim at describing the natural variability of a shape, e.g. the morphological variation of the same bone from different subjects or the organ of one subject through time. The general idea behind SSM is to perform a linear decomposition of the shape variability from a set of training data by defining a mean shape and modes of deformations under some mathematical criteria. The most commonly used principal component analysis (PCA) estimates orthogonal directions which maximize the variance. Other linear decompositions can be used, based on principal factor analysis or independent component analysis. PCA is mathematically defined as an orthogonal linear transformation that transforms the data to a new coordinate system such that the greatest variance by any projection of the data comes to lie on the first axis (called the first principal component), the second greatest variance on the second axis, and the third greatest variance on the third axis and so on. PCA is theoretically the optimum transform for a given data in the least squares sense, its operation can be thought of as revealing the internal structure of the data in a way that best explains the variance in the data (Dryden and Mardia 1998; Jolliffe 2002). PCA-based classification methods implicitly assume that the data was generated by a Gaussian probability distribution and is therefore less than optimal if this assumption is not valid.

Shape descriptors used in medical image analysis include both parametric models, such as Fourier descriptors or spherical harmonic functions, and non-parametric models such as landmark-based descriptors, deformation fields, distance transforms and medial axes. We explore landmark-based descriptors or more specifically shape analysis based on densely sampled 3D point distribution models. Landmarks can be generally grouped as:

Anatomical landmarks: Points assigned by an expert or automatically extracted that have a correspondence between organisms in some biologically meaningful way.

Mathematical landmarks: Points located on an object according to some mathematical or geometrical property, e.g. centre of a ball and socket joint obtained through least squares fitting of a sphere.

Pseudo-landmarks: Constructed points on an object either around the outline or between anatomical landmarks.

In general, a statistical shape model is built from a training set of shapes $\{\mathrm{Shape_i}: i = 1, \ldots, n_s\}$ where n_s is the total number of shapes. Each shape can be approximated by $n_p$points sampled on its 3D surface such that the individual points are placed in corresponding positions across the set of shapes. This results in a set of initial shape vectors $\{S_i : i = 1, \ldots, n_s\}$ which form a representation of the whole set of shapes in a common shape space $\mathbb{R}^{3\times n_p}$.

Using PCA, each shape vector can be expressed using a linear model (Eq. 1), that is, new shape instances can be obtained by adding a linear combination of the principal shape modes to the mean surface:

$$S = \bar{S} + Pb = \bar{S} + \sum\nolimits_{i=1}^{n_s-1} p_i b_i \tag{1}$$

where $\bar{S} = \frac{1}{n_s} \sum\nolimits_{i=1}^{n_s} S_i$ is the mean shape vector and $P = \{p_i\}$ the matrix of eigenvectors of the covariance matrix. The corresponding eigenvalues $\{\lambda_i\}$ describe the amount of variance in the direction of the eigenvectors, in practice, what is retained is not simply the directions corresponding to non-zero eigenvalues, but instead that ordered set, from highest to lowest, which adds to a certain amount of the total variance of the data, where the total variance of the data is given by $\sum\nolimits_{i=1}^{n_s-1} \lambda_i$, and the variance up to n_m is $\sum\nolimits_{i=1}^{n_m} \lambda_i$, here the number of modes n_m is chosen to be the lowest value such that the cumulative variance up to n_m is above a certain threshold (e.g. 99%). This is a somewhat arbitrary value allowing for a compact model whilst capturing most of the variance in the data. Other commonly used values are 90% and 95%. The shape parameters $b = \{b_i\}$ control the contribution of modes of variation, $b_i$is usually chosen to be bounded between $-3\sqrt{\lambda_i}$ and $+3\sqrt{\lambda_i}$.

In order to obtain a correct statistical model all points on each surface must correspond in an anatomical meaningful way, and their coordinates must be given relative to a common frame of reference (alignment). These two steps are fundamental, since incorrect correspondences can either introduce too much variation or lead to erroneous instances of the model (Dryden and Mardia 1998; Lamecker et al. 2004).

2.2 *Alignment and Non-rigid Registration*

Before establishing correspondences using non-rigid surface registration, alignment and scaling can be achieved using the automatically extracted anatomical landmark points and an iterative approach using Procrustes analysis (Bookstein 1997; Cootes et al. 2004).

The non-rigid registration of the bones is based on the method introduced by Rueckert et al. 1999. The key idea of this method is to embed a surface in a volumetric mesh which defines the continuous deformation field. The mesh is subsequently subdivided into higher resolution levels by inserting control points into the current level of control points and decreasing the mesh space. The deformation is affected by the spacing of the control points: a larger spacing leads to a smoother global shape while a relatively smaller spacing leads to higher local deformation. It is based on a Free Form Deformation (FFD) with a B-spline grid, and error calculation is done with normalized mutual information. To define a spline based FFD, we denote the domain of the bone volume as $\Omega = \{(x; y; z)| 0 \leq x < X, 0 \leq y < Y, 0 \leq z < Z\}$.

Let Φ denote a $nx \times ny \times nz$ mesh of control points $\Phi(i,j,k)$with uniform spacing dx,dy and dz. Then, the FFD with transformation T can be written as the 3-D tensor product of the 1D cubic B-splines (Eqs. 2, 3, and 4)

$$T = \sum_{l=0}^{3}\sum_{m=0}^{3}\sum_{n=0}^{3} B_l(u)B_m(v)B_n(w)\Phi(i+l, j+m, k+n) \tag{2}$$

where

$$\begin{aligned} &i = \lfloor x/nx \rfloor - 1, \quad j = \lfloor y/ny \rfloor - 1, \quad k = \lfloor z/nz \rfloor - 1, \\ &u = x/dx - \lfloor x/dx \rfloor, \; v = y/dy - \lfloor y/dy \rfloor, \; w = z/dz - \lfloor z/dz \rfloor \end{aligned} \tag{3}$$

$\boldsymbol{B_l}$ represents the lth basis function of the B-spline

$$\begin{aligned} B_0(u) &= (1-u)^3/6 \\ B_1(u) &= (3u^3 - 6u^2 + 4)/6 \\ B_2(u) &= (-3u^3 - 3u^2 + 3u + 1)/6 \\ B_3(u) &= u^3/6 \end{aligned} \tag{4}$$

2.3 *Illustrative Results: Femur Bone Main Modes of Shape Variation*

Compactness of an SSM of the femur, based on 43 femurs (Fig. 3). The first mode explains approximately 48% of the total observed shape variance, the seven first combined modes account for about 90% and the 20 first modes capture more than 99% of shape variation.

In Fig. 4, the first three modes of shape variance and mean shape can be visualized.

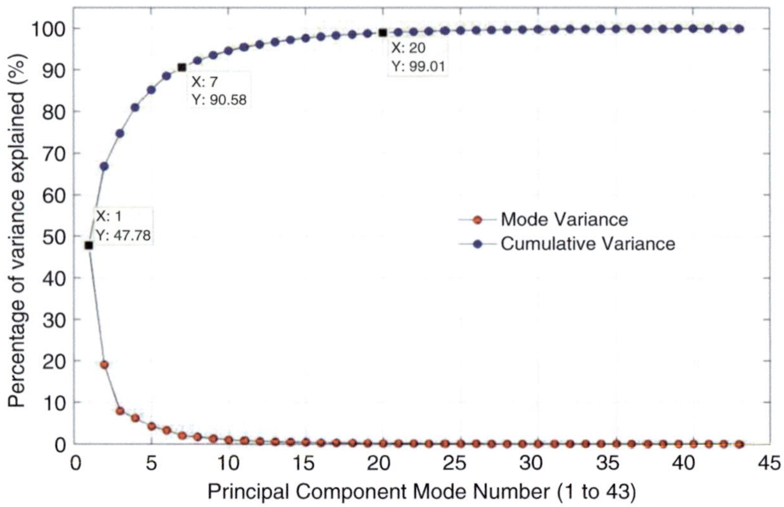

Fig. 3 Compactness of an SSM of the femur, based on 43 bones

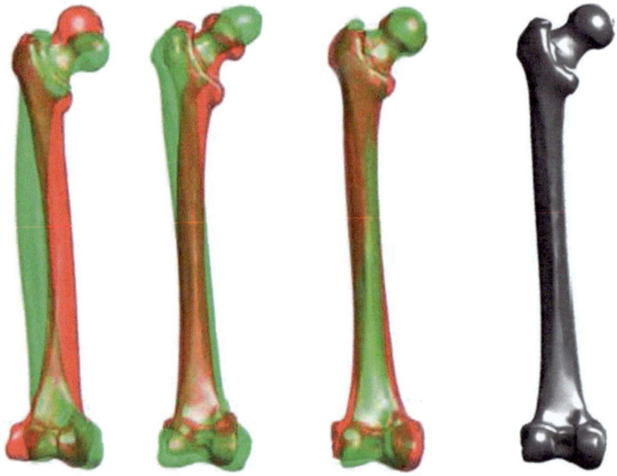

Fig. 4 Mean shape and variations of the shape of the full femur for the first three modes, based on an SSM built from 43 training femurs. From *left* to *right* modes one to three and mean shape (*grey*), green using a shape parameter of $-3\sqrt{\lambda_i}$ and *red* $+3\sqrt{\lambda_i}$

3 Integration of Virtual Planning into Computer-Aided Surgery

Diagnosis and correction of osseous deformity and soft tissue function are among the most common aspects of orthopedic surgery. Commercial and open- source computer aided-tools aiming at virtual analysis and simulation have increasingly began to emerge parallel to the technical improvements in imaging techniques, allowing for detailed volumetric visualization of the musculoskeletal system

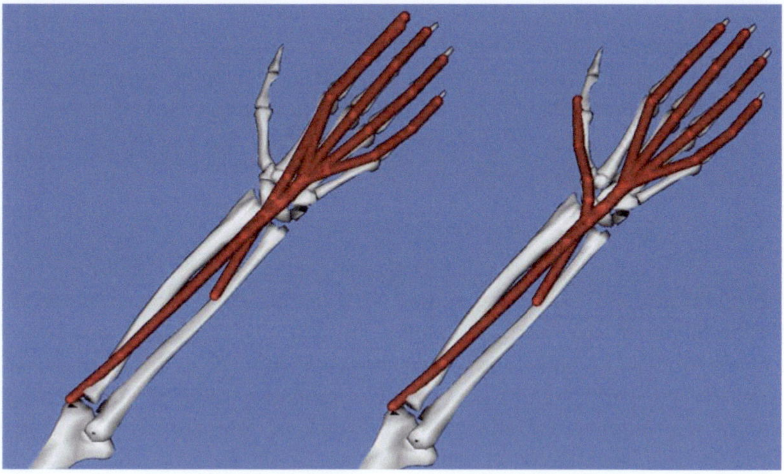

Fig. 5 OpenSim® wrist model simulating an extensor indices proporius to extensor pollicus longus transfer

(Blemker et al. 2007; Delp et al. 2007; Delp and Loan 1995). Techniques such as CT and MR imaging are currently routinely available. The goals of computer-based surgery simulation are to enable the surgeon to experiment with different surgical procedures (osteotomies, grafts, implants, etc.) in an artificial environment and to try predicting the outcome of an intervention before the actual surgery. However, the final surgical outcome and accuracy in implementing the virtual plan are related to the surgeon's ability to reproduce the planned surgery.

One of today's state of the art examples of simulation software is OpenSim, an open-source software system developed at Stanford University that lets users create and analyze dynamic simulations of movement (Delp et al. 2007). It can be used to study common orthopedic surgical techniques such as muscle transfers, and allows for analysis of how muscle moment arm, optimal muscle fiber length, and tendon slack length influence the variation of muscle force with respect to changes in joint angle. Customized models can be created by scaling of a generic model (Fig. 5).

Despite the growing dissemination and technical excellence of these virtual modeling tools, implementation of a virtual plan in the operating theatre and the linkage to the available navigational systems remain a restraint to full computer aided assistance, from image to actual surgery. Part of the problem lies in the fact that both academic researchers and industry, each in their individual attempt to attain international excellence, have been focusing on and specializing in subdisciplines such as image analysis, modeling, surgical robotics or navigation with limited integration possibilities between each.

Currently, "integrated and multiscale approaches" have become important research goals in ICT for health. A promising example is the recently granted European Project Tlemsafe, joining academic experts in image analysis and segmentation, musculoskeletal modeling and navigation technology, and including industrial partners such as Materialise Inc (Image analysis and rapid prototyping), Anybody Inc (Muscoloskeletal modeling) and Brainlab Inc (surgical navigation).

To date, a few methods are already readily available to make the integration from image over virtual planning to computer-assisted surgery. We present two such solutions, either using customized surgical guides or by generating a virtual plan in DICOM format that can be read by common navigation systems.

3.1 Patient-Specific Surgical Guides

In the orthopedic field, surgical standard instruments have been used for years to transfer the planning into surgical gestures. However, the exact position of standard instruments relative to patient anatomy was not defined, and had to be adapted intraoperatively in order to reproduce the planning as well as possible. As opposed to standard instruments, patient-specific guides, or templates/jigs, started to be developed in the late 1990s (Birnbaum et al. 2001; Radermacher et al. 1998). These surgical guides generally possess a surface that exactly fits a patient's anatomical surface, e.g. bone or cartilage surface. During surgery, the guide is positioned onto this anatomical surface and its position tested until the optimal match between guide and patient anatomical surface is found, which indicates that the guide is correctly positioned thanks to a unique fitting position.

While standard instruments are built using classical manufacturing techniques, e.g. injection molding, the use of similar techniques for building patient-specific guides would be extremely cost-inefficient. This explains why the first patient-specific guides were built using a milling technique, starting from a standardized block that was progressively milled to reach the final shape of the surgical guide (Radermacher et al. 1998). An interesting feature of this technique is the existence of desktop milling machines. The design and production of these guides were thus presented as a future integrated process in the medical routine and performed by the medical staff.

Another means for building patient-specific guides are the rapid prototyping, or additive manufacturing, techniques (Hopkinson et al. 2005). The basic principles of these techniques consist of, starting from a virtual object to be produced, virtually slicing it into a stack of layers. The object is then built layer per layer. Different materials can be used for rapid prototyping objects, and in case of surgical guides, the selective laser sintering technique based on polyamide powder is often used, mainly because of its manufacturing accuracy, biocompatibility, and good response to heating for sterilization. Main advantages of rapid prototyping techniques for surgical guides are that any shape can be built and that patient-specificity of guides does not imply any additional cost. This explains why nowadays, most of the patient-specific surgical guides are built using these techniques. Interestingly, rapid prototyping techniques have also been used to build bone models for improving accuracy and efficiency of surgeries by helping with visualization, understanding, and planning of cases (Bagaria et al. 2011; Brown et al. 2003), as well as by pre-forming implants (Ahn et al. 2006).

Patient-specific surgical guides for orthopedic applications were initially developed for spine surgeries and pelvic osteotomies (Radermacher et al. 1998). Since then, the number of applications has grown drastically, both in academic and industrial

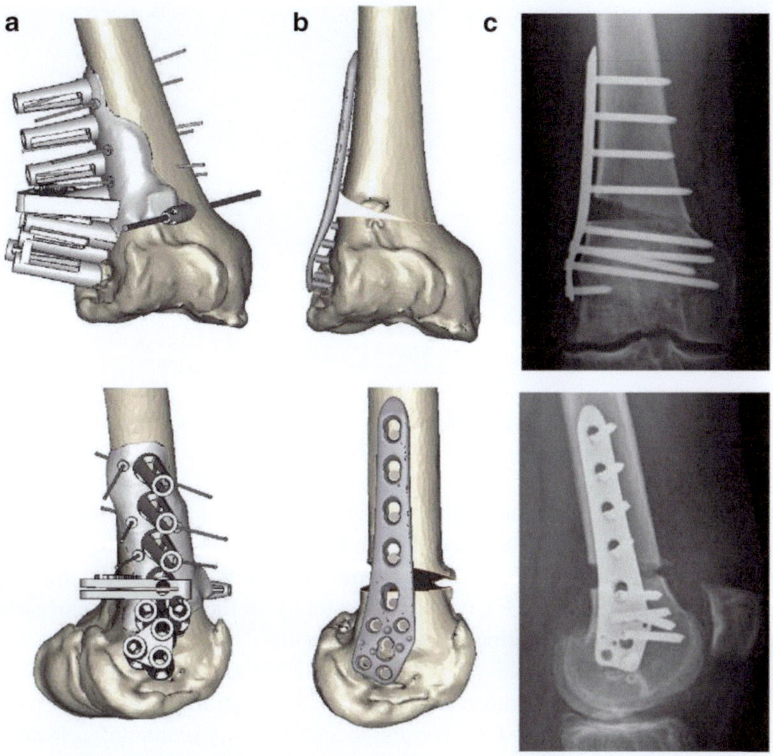

Fig. 6 Illustration of a patient-specific guide for distal femur osteotomy. (**a**) Guide fit on femur bone model. (**b**) Planning of surgery. (**c**) Postoperative X-rays

environments. One can now find patient-specific guides for various surgical procedures such as total knee replacement (Hafez et al. 2006; Lombardi et al. 2008), total hip replacement (Hananouchi et al. 2009), hip surface replacement (Pattyn et al. 2010; Raaijmaakers et al. 2010; Kunz et al. 2010), intra- and extra-articular radius malunions (Leong et al. 2010; Oka et al. 2008), and distal femur osteotomy (Victor et al. 2011), Fig. 6.

For each application, a guide can provide different types of guidance. It can mechanically guide surgical gestures, such as drilling or cutting (Pattyn et al. 2010; Raaijmaakers et al. 2010; Kunz et al. 2010; Leong et al. 2010; Oka et al. 2008; Victor et al. 2011) or indicate anatomical landmarks for positioning standard instruments (Blanz et al. 2004). It can guide the placement of a plate by providing the drilling positions and orientations of plate fixations on preoperative bone (Leong et al. 2010; Victor et al. 2011). It can also visually guide a surgical gesture such as the impaction of a hip implant (Blemker et al. 2007). It can fully replace a standard instrument, becoming for example a patient-specific cutting block for total knee replacement (Hafez et al. 2006). Finally, it can be left in the body, becoming a guiding implant to reposition bone fragments for periacetabular osteotomy (Anaya et al. 2011) or distal radius osteotomy (Oka et al. 2010).

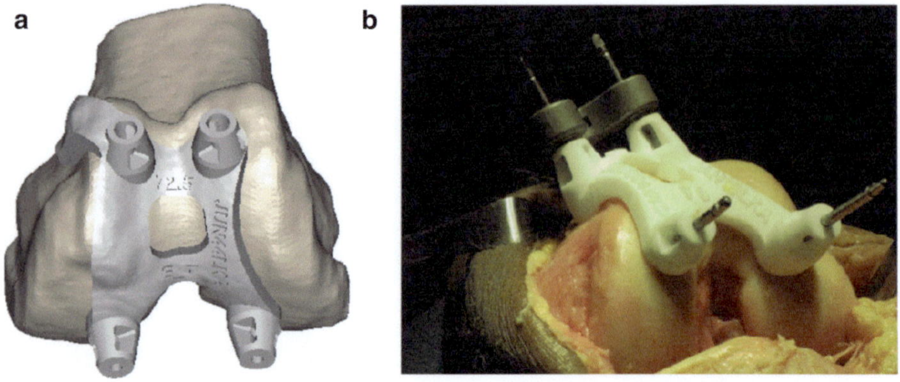

Fig. 7 Illustration of patient-specific Signature™ femoral guide for total knee replacement. (**a**) Guide fit on femur bone model. (**b**) Use of guide during surgery

Patient-specific surgical guides were initially developed for increasing the accuracy of procedures (Radermacher et al. 1998). For numerous applications, the goal for using guides is still similar. However, this is not the goal pursued by the use of guides for total knee replacement (TKR), which has by far made the biggest progression. This is demonstrated by the routine use of Signature™ (Biomet), Patient Specific Instruments ®(Zimmer), OtisKnee™ (OtisMed), VISIONAIRE™ (Smith & Nephew), PROPHECY™ (Wright) or TruMatch™ (Depuy) for high number of surgeries. The main goal there is to reduce the OR time and to avoid using intramedullar instruments, which also decreases the risk of infection (Hafez et al. 2006). Without the use of a CAS system, a TKR procedure is generally performed in three steps: imaging (e.g. X-rays), planning of the procedure to define the approximate position and size of the implant, and finally surgery itself. During surgery, intramedullar rods and sets of gauges are used to correctly position the cutting blocks on the femur and tibia sides. When using a patient-specific surgical guide such as the Signature™ guide (Fig. 7), the TKR procedure is now performed in five steps. It starts by the imaging of the patient. 3D images (including parts of hip and ankle), such as CT or MRI, are acquired, which is necessary for the subsequent planning and design of the guide. Knee bones, and possibly cartilage, are then segmented from the 3D image. The planning of the TKR consists in defining the position and orientations of the cuts, in addition to the position and the size of the implant. The planning is performed in 3D by calculating the mechanical axes using 3D femur and tibia models. Based on the planned cuts, the position of cutting blocks can be deducted. The surgical guides are then designed such that fixation pins of the cutting blocks are drilled through the guides. During surgery, all the surgical gestures and instruments that are used for deriving the fixation of the cutting blocks can thus be discarded.

Although there is an increasing use of patient-specific surgical guides, interesting challenges remain to make its growth even bigger. Using guides not only requires the acquisition of 3D images, but also a planning performed in 3D. Evolution of the planning, going from 2D on X-rays (as previously often performed) to 3D on CT/MRI, will thus be necessary for each procedure. When using surgical guides, going from image acquisition to the surgery itself is lengthened by the 3D segmentation of bones, the 3D planning, the design and production of guides. So far, the lead time for the surgeon to receive the surgical guides remains too long to allow using patient-specific guides for trauma surgeries. Automation of several steps is thus crucial for guides to be used during these surgeries. Finally, most of the patient-specific guides are still not adjustable intraoperatively. Although their use requires thorough planning, which thus implies a lower risk of need for adaptation during surgery, one cannot entirely discard this need. Besides, as explained above, the lead time is still too long to provide new guides rapidly. The development of patient-specific adjustable guides should thus be also envisaged.

3.2 Computer-Assisted Surgical Planning and Navigation

Surgical planning for osteotomies, tumor resection, implant positioning and deformity correction or bone realignment can easily be performed in 3D before any actual surgery is being performed. Image processing software such as Mimics® (Materialise NV, Heverlee, Belgium) allows for the creation of detailed 3D virtual models from MRI and CT modalities, which can be cut, non-uniformly scaled and oriented in 6 degrees of freedom. In addition and of interest in tumor surgery, data from different imaging modalities can be fused into a single anatomical model (Wong et al. 2010). This is of particular interest in oncology cases for the analysis of tumor extension and its relation with regional anatomy. Fusion of CT and MRI yields hybrid images that combine the key characteristics of each technique, enabling better interpretation and localization of the lesion. The virtual plans can further be exported in DICOM format and used in almost any navigation software (Wong et al. 2008).

Alternatively, and for more complex planning, mesh files such as the STL format can be generated from the patient imaging files and used in a musculoskeletal model package allowing for dynamic simulations and evaluation of "what if" scenarios, to define an optimized surgical plan. Once an optimized virtual plan has been achieved, the resulting data can be re-integrated into the patient's original imaging data set. Again, the resulting data can be exported in DICOM format for further use in computer-assisted surgical navigation (Audenaert et al. 2011).

An example of this approach has been used at our orthopedic department for femoroacetabular impingement (FAI). FAI is a mechanical hip disorder defined as early and/or repetitive contact between osseous prominences of the acetabulum and/or the proximal femur, potentially resulting in damage to the hip joint in young adults. The prevalence of these lesions – including asymptomatic cases – has been reported to be as high as 17–25%, and multiple studies have shown a significant relationship between radiographic parameters specific of FAI and the development of osteoarthritis of the hip. It has been estimated that osteoarthritis of the hip in Caucasian patients is actually secondary to FAI in 40–70% of cases. Computer-aided technologies can aid in the early diagnosis and adequate treatment of FAI and

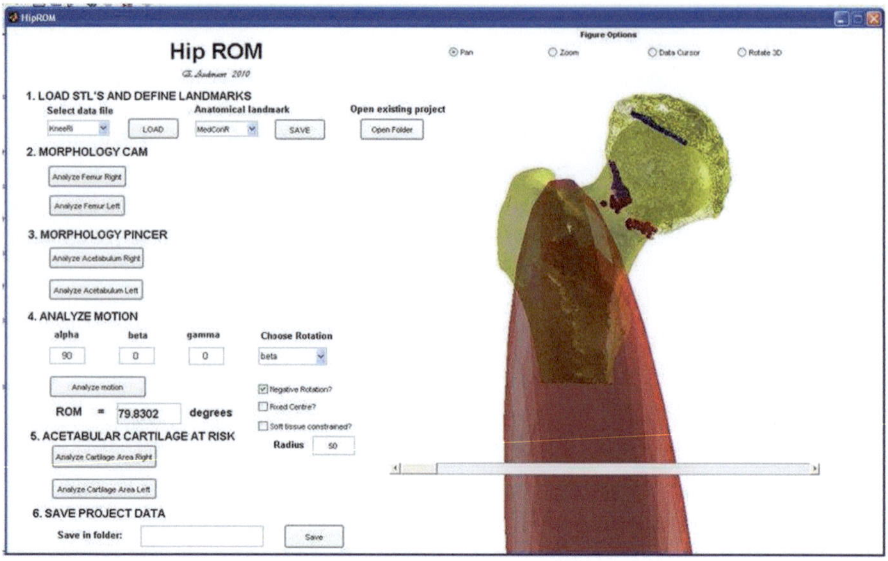

Fig. 8 Patient-specific virtual planning and optimization of treatment

thereby prevent or slow down the progression to osteoarthritis. In addition, surgical treatment is technically demanding and involves a steep learning curve. Computer-aided surgery and virtual surgical planning and training can help junior surgeons and surgeons in training to overcome these difficulties and thereby increase the rate of successful surgeries.

Patients are scanned using a 64-slice CT scan (LightSpeed VCT, GE Healthcare, Milwaukee, Wis). A density-based automated segmentation of the femur and pelvis is then performed using the Mimics® image processing software package. Patient-specific STL-format files of the bones are generated and exported for further analysis outside the image analysis software. A custom-designed package is then used for kinematic analysis of the patient's hip joint. The package provides a graphical user interface that allows for morphological evaluation of bony anomalies as well as for simulation of movement and range of motion by collision detection (Audenaert et al. 2010, 2011). The necessary amount of bone resection to correct anatomy and restore normal ROM can then be defined. In addition, the effect of possible confounding parameters, e.g. hip joint centre position or soft tissue restraints such as muscles or ligaments impeding motion, can be analyzed (Fig. 8).

A virtual plan of the optimized surgical correction to be obtained is exported as STL-format and matched to the original patient imaging data. The virtual plan is then exported in DICOM format for use by the navigation system. In our case the Stryker Orthomap 3D navigation and planning software (Stryker, Mahwah, NJ, USA) is used. The Stryker OrthoMap 3D navigation system requires DICOM image sets as a basis for navigation. Their planning software provides segmentation tools that allow the user to accentuate anatomical structures of interest, such as tumours, bone surfaces and vessels. In our case, a cam and/or pincer lesion is accentuated. The virtual models are matched with the patient's anatomy preoperatively by using

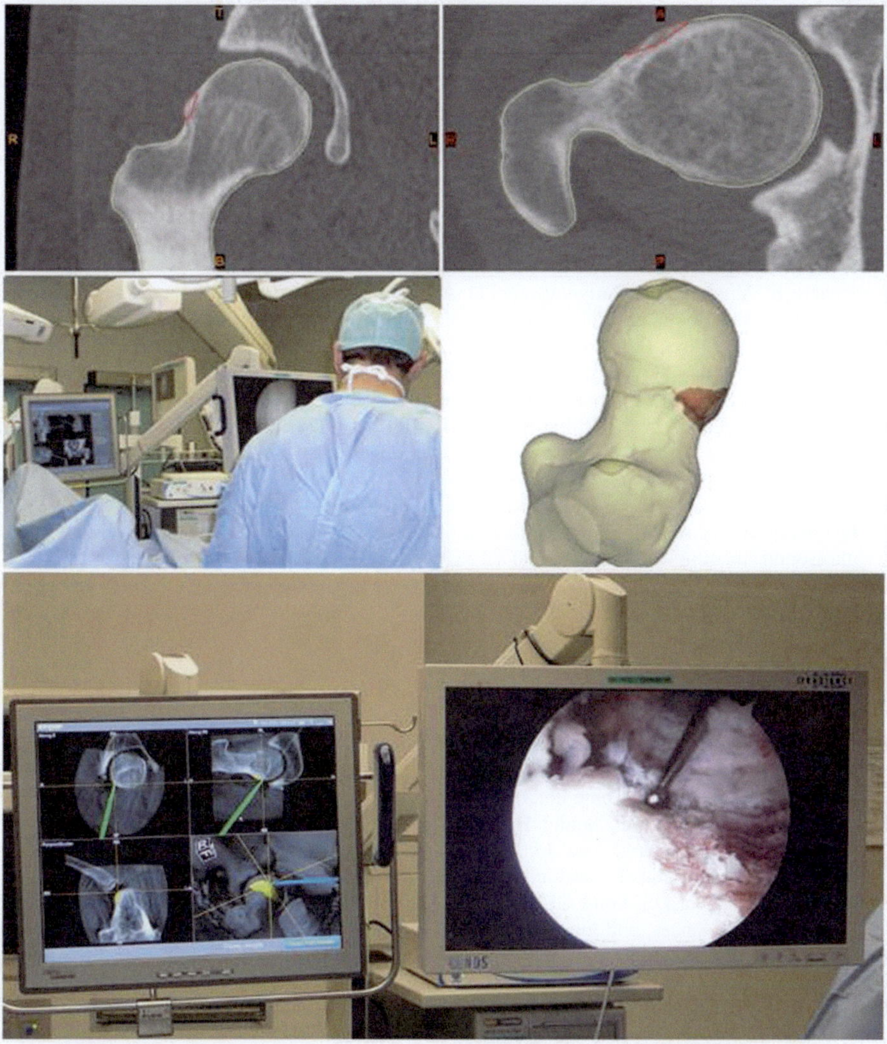

Fig. 9 Clinical case of femoroacetabular impingement. From image to virtual plan and image-based computer assisted surgery

a surface-to-surface matching algorithm. Both image-free and image-based navigation protocols can theoretically be used. However, we found significant accuracy problems during the registration process when applying the image-free protocol. The minimal invasive character of the arthroscopic procedure limits access to the relevant bony structures. This impedes accurate surface-to-surface matching. In case of the image-based protocol, intra-operative matching of the patient anatomy was obtained using the Ziehm Vision 3D fluoroscope (Zhiem, Nürnberg, Germany).

Finally, a universal tracker can be attached to a calibrated surgical drill and the lesion is resected under computer guidance (Fig. 9).

References

Ahn DG, Lee JY, Yang DY (2006) Rapid prototyping and reverse engineering application for orthopedic surgery planning. J Mech Sci Technol 20(1):19–28

Anaya AM, Vigneron L, Diab M, Burch S (2011) Evaluation of virtual planning, rapid prototyping modeling, and image-guided navigation in periacetabular osteotomy. In: Proceedings of computer assisted radiology and surgery 2011, S111

Audenaert EA, Baelde N, Huysse W, Vigneron L, Pattyn C (2010) Development of a three-dimensional detection method of cam deformities in femoroacetabular impingement. Skeletal Radiol 40:921–927

Audenaert E, Vigneron L, Pattyn C (2011) A method for three-dimensional evaluation and computer aided treatment of femoroacetabular impingement. Comput Aided Surg 16:143–148

Bagaria V, Deshpande S, Rasalkar DD, Kuthe A, Paunipagar BK (2011) Use of rapid prototyping and three-dimensional reconstruction modeling in the management of complex fractures. Eur J Radiol 80(3):814–820. Epub 2011 Jan 22. Review. http:/pubmed/21256690

Baldwin MA, Langenderfer JE, Rullkoetter PJ, Laz PJ (2010) Development of subject specific and statistical shape models of the knee using an efficient segmentation and mesh-morphing approach. Comput Methods Programs Biomed 97:232–240

Barratt DC, Chan CS, Edwards PJ, Penney GP, Slomczykowski M, Carter TJ et al (2008) Instantiation and registration of statistical shape models of the femur and pelvis using 3D ultrasound imaging. Med Image Anal 12:358–374

Behiels G, Maes F, Vandermeulen D, Suetens P (2002) Evaluation of image features and search strategies for segmentation of bone structures in radiographs using active shape models. Med Image Anal 6:47–62

Benameur S, Mignotte M, Parent S, Labelle H, Skalli W, de Guise J (2003) 3D/2D registration and segmentation of scoliotic vertebrae using statistical models. Comput Med Imag Graph 27:321–337

Birnbaum K, Schkommodau E, Decker N, Prescher A, Klapper U, Radermacher K (2001) Computer-assisted orthopedic surgery with individual templates and comparison to conventional operation method. Spine 26:365–370

Blanz V, Mehl A, Vetter T, Seidel HP. (2004) A statistical method for robust 3D surface reconstruction from sparse data. In: International symposium on 3D data processing, visualization and transmission, Thessaloniki, Greece, pp 293–300

Blemker SS, Asakawa DS, Gold GE, Delp SL (2007) Image-based musculoskeletal modeling: applications, advances, and future opportunities. J Magn Reson Imaging 25(2):441–451

Bookstein FL (1997) Landmark methods for forms without landmarks: morphometrics of group differences in outline shape. Med Image Anal 1(3):225–243

Brown GA, Firoozbakhsh K, DeCoster TA, Reyna JR Jr, Moneim M (2003) Rapid prototyping: the future of trauma surgery? J Bone Joint Surg 85(4):49–55

Bryan R, Nair PB, Taylor M (2009) Use of a statistical model of the whole femur in a large scale, multi-model study of femoral neck fracture risk. J Biomech 42:2171–2176

Cerveri P, Marchente M, Bartels W, Corten K, Simon JP, Manzotti A (2010) Automated method for computing the morphological and clinical parameters of the proximal femur using heuristic modeling techniques. Ann Biomed Eng 38(5):1752–1766

Cootes TF, Taylor CJ (2004) Anatomical statistical models and their role in feature extraction. Br J Radiol 77:S133–S139

Cootes TF, Taylor CJ, Cooper D, Graham J (1995) Active shape models – their training and application. Comput Vis Image Underst 61:38–59

Cootes TF, Taylor CJ, Manchester M Pt (2004) Statistical models of appearance for computer vision. (Online: http://citeseerx.ist.psu.edu/viewdoc/summary?doi = 10.1.1.58.1455)

Delaunay S, Dussault RG, Kaplan PA, Alford BA (1997) Radiographic measurements of dysplastic adult hips. Skeletal Radiol 26(2):75–81

Delp SL, Loan JP (1995) A graphics-based software system to develop and analyze models of musculoskeletal structures. Comput Biol Med 25(1):21–34

Delp SL, Anderson FC, Arnold AS, Loan P, Habib A, John CT, Guendelman E, Thelen DG (2007) OpenSim: open-source software to create and analyze dynamic simulations of movement. IEEE Trans Biomed Eng 54:1940–1950

Dryden IL, Mardia K (1998) Statistical shape analysis. Wiley, Chichester

Ehrhardt J, Handels H, Plötz W, Pöppl SJ (2004) Atlas-based recognition of anatomical structures and landmarks and the automatic computation of orthopedic parameters. Methods Inf Med 43 (4):391–397

Fleute M, Lavallée S (1998) Building a complete surface model from sparse data using statistical shape model. Lect Notes Comput Sci 1496:879–887

Gomes GT (2011) Automatic feature extraction and statistical shape analysis of the Femur. Submitted to computer methods in biomechanics and biomedical engineering

Grood ES, Suntay WJ (1983) A joint coordinate system for the clinical description of three-dimensional motions: application to the knee. J Biomech Eng 105(2):136–144

Hafez MA, Chelule KL, Seedhom BB, Sherman KP (2006) Computer-assisted total knee arthroplasty using patient-specific templating. Clin Orthop Relat Res 444:184–192

Hananouchi T, Saito M, Koyama T, Hagio K, Murase T, Sugano N, Yoshikawa H (2009) Tailor-made surgical guide based on rapid prototyping technique for cup insertion in total hip arthroplasty. Int J Med Robot 5(2):164–169

Hankemeier S, Gosling T, Richter M, Hufner T, Hochhausen C, Krettek C (2006) Computer-assisted analysis of lower limb geometry: higher intraobserver reliability compared to conventional method. Comput Aided Surg 11(2):81–86

Heimann T, Meinzer HP (2009) Statistical shape models for 3D medical image segmentation: a review. Med Image Anal 13:543–563

Hopkinson N, Hague R, Dickens P (2005) Rapid manufacturing: an industrial revolution for the digital age. Wiley, Chichester

Jolliffe IT (2002) Principal component analysis, 2nd edn. Springer, New York

Kunz M, Rudan JF, Xenoyannis GL, Ellis RE (2010) Computer assisted hip resurfacing using individualized drill templates. J Arthroplasty 25(4):600–606

Kurazume R, Nakamura K, Okada T, Sato Y, Sugano N, Koyama T, Iwashita Y, Hasegawa T (2009) 3D reconstruction of a femoral shape using a parametric model and two 2d fluoroscopic images. Comput Vis Image Underst 113(2):202–211

Lamecker H, Seebass M, Hege HC, Deuflhard P (2004) A 3d statistical shape model of the pelvic bone for segmentation. In: Fitzpatrick JM, Sonka M (eds), SPIE 5370(1):1341–1351

Leong NL, Buijze GA, Fu EC, Stockmans F, Jupiter JB (2010) Computer-assisted versus non-computer-assisted preoperative planning of corrective osteotomy for extra-articular distal radius malunions: a randomized controlled trial. BMC Musculoskelet Disord 11:282

Lerner AL, Tamez-Pena JG, Houck JR, Yao J, Harmon HL, Salo AD, Totterman SM (2003) The use of sequential MR image sets for determining tibiofemoral motion: reliability of coordinate systems and accuracy of motion tracking algorithm. J Biomech Eng 125(2):246–253

Lombardi AV, Berend KR, Adams JB (2008) Patient-specific approach in total knee arthroplasty. Orthopedics 31(9):927–930

Morton NA, Maletsky LP, Pal S, Laz PJ (2007) Effect of variability in anatomical landmark location on knee kinematic description. J Orthop Res 25(9):1221–1230

Nizard R (2002) Computer assisted surgery for total knee arthroplasty. Acta Orthop Belg 68 (3):215–230

Nofrini L, Slomczykowski M, Iacono F, Marcacci M (2004) Evaluation of accuracy in ankle center location for tibial mechanical axis identification. J Invest Surg 17(1):23–29

Oka K, Moritomo H, Goto A et al (2008) Corrective osteotomy for malunited intra-articular fracture of the distal radius using a custom-made surgical guide based on three-dimensional computer simulation: case report. J Hand Surg Am 33:835–840

Oka K, Murase T, Moritomo H, Goto A, Sugamoto K, Yoshikawa H (2010) Corrective osteotomy using customized hydroxyapatite implants prepared by preoperative computer simulation. Int J Med Robot 6(2):186–193

Paley D (2002) Normal lower limb alignment and joint orientation. In: Paley D, Herzenberg JE (eds) Principles of deformity correction. Springer, New York

Pattyn C, De Smedt K, Gelaude F, Clijmans T, Dille J, Geebelen B, Audenaert E (2010) A custom-made guide for femoral component positioning in hip resurfacing arthroplasty: development and validation study. J Biomech 43(1)

Raaijmaakers M, Gelaude F, De Smedt K, Clijmans T, Dille J, Mulier M (2010) A custom-made guide-wire positioning device for hip surface replacement arthroplasty: description and first results. BMC Musculoskelet Disord 11:161

Radermacher K, Portheine F, Anton M, Zimolong A, Kaspers G, Rau G, Staudte HW (1998) Computer assisted orthopaedic surgery with image-based individual templates. Clin Orthop Relat Res 354:28–38

Rajamani KT, Styner MA, Talib H, Zheng G, Nolte LP, Gonzáles Ballester MA (2007) Statistical deformable bone models for robust 3D surface extrapolation from sparse data. Med Image Anal 11:99–109

Rueckert D, Sonoda LI, Hayes C, Hill DLG, Leach MO, Hawkes DJ (1999) Non-rigid registration using free-form deformations: application to breast MR images. IEEE Trans Med Imag 18:712–721

Schöttle PB, Schmeling A, Rosenstiel N, Weiler A (2007) Radiographic land-marks for femoral tunnel placement in medial patellofemoral ligament reconstruction. Am J Sports Med 35 (5):801–804

Siston RA, Giori NJ, Goodman SB, Delp SL (2007) Surgical navigation for total knee arthroplasty: a perspective. J Biomech 40(4):728–735

Stindel E, Briard JL, Merloz P, Plaweski S, Dubrana F, Lefevre C et al (2002) Bone morphing: 3D morphological data for total knee arthroplasty. Comput Aid Surg 7:156–168

Styner M, Lieberman JA, McClure RK, Weinberger DR, Jones DW, Gerig G (2005) Morphometric analysis of lateral ventricles in schizophrenia and healthy controls regarding genetic and disease specific factors. Proc Natl Acad Sci USA 102(13):4872–4877

Subburaj K, Ravi B, Agarwal M (2009) Automated identification of anatomical landmarks on 3D bone models reconstructed from CT scan images. Comput Med Imaging Graph 33(5):359–368

Subburaj K, Ravi B, Agarwal M (2010) Computer-aided methods for assessing lower limb deformities in orthopaedic surgery planning. Comput Med Imaging Graph 34(4):277–288

Tang TS, Ellis RE (2005) 2D/3D deformable registration using a hybrid atlas. Med Image Comput Comput Assist Interv 8:223–230

Van Cauter S, De Beule M, Van Haver A, Verdonk P, Verhegghe B (2012) Automated extraction of the femoral anatomical axis for determining the intramedullary rod parameters in total knee arthroplasty. Int J Numer Methods Biomed Eng 28(1):158–169

Van Sint Jan S (2007) Color atlas of skeletal landmark definitions. Guidelines for reproducible manual and virtual palpations. Churchill Livingstone–Elsevier, Edinburgh

Van Sint Jan S, Della Croce U (2005) Identifying the location of human skeletal landmarks: why standardized definitions are necessary-a proposal. Clin Biomech (Bristol, Avon) 20(6):659–660

Victor J, Van Doninck D, Labey L, Innocenti B, Parizel PM, Bellemans J (2009) How precise can bony landmarks be determined on a CT scan of the knee? Knee 16(5):358–365

Victor J, Deprez P, Premanathan A, Keppler L (2011) Virtual 3d planning and patient specific surgical guides for osteotomies around the knee. In: Proceedings of Computer Assisted Orthopaedics Surgery. Digital Proceedings so far (CD-ROM). They will be published in JBJS: http://www.bjjprocs.boneandjoint.org.uk/site/includefiles/abstractspending.xhtml

Wolf A, Digioia AM 3rd, Mor AB, Jaramaz B (2005) Cup alignment error model for total hip arthroplasty. Clin Orthop Relat Res 437:132–137

Wong KC, Kumta SM, Antonio GE, Tse LF (2008) Image fusion for computer-assisted bone tumor surgery. Clin Orthop Relat Res 466(10):2533–2541

Wong KC, Kumta SM, Leung KS, Ng KW, Ng EW, Lee KS (2010) Integration of CAD/CAM planning into computer assisted orthopaedic surgery. Comput Aided Surg 15(4–6):65–74

Yang YM, Rueckert D, Bull AMJ (2008) Predicting the shapes of bones at a joint: application to the shoulder. Comput Methods Biomech Biomed Eng 11(1):19–30

Yoon YS, Hodgson AJ, Tonetti J, Masri BA, Duncan CP (2008) Resolving inconsistencies in defining the target orientation for the acetabular cup angles in total hip arthroplasty. Clin Biomech 23(3):253–259

Yoshino N, Takai S, Ohtsuki Y, Hirasawa Y (2001) Computed tomography measurement of the surgical and clinical transepicondylar axis of the distal femur in osteoarthritic knees. J Arthroplasty 16(4):493–497

Zheng G, Schumann S (2009) 3D reconstruction of a patient-specific surface model of the proximal femur from calibrated X-ray radiographs: a validation study. Med Phys 36:1155–1166

Zheng G, Gonzáles-Ballester MA, Styner M, Nolte LP (2006) Reconstruction of patient specific 3d bone surface from 2d calibrated fluoroscopic images and point distribution model. Lect Notes Comput Sci 4190:25–32

Zheng G, Gollmer S, Schumann S, Dong X, Feilkas T, Ballester MAG (2008) A 2d/3d correspondence building method for reconstruction of a patient-specific 3d bone surface model using point distribution models and calibrated X-ray images. Med Image Anal 13(6):883–899

Ziegler CG, Pietrini SD, Westerhaus BD, Anderson CJ, Wijdicks CA, Johansen S, Engebretsen L, Laprade RF (2010) Arthroscopically pertinent landmarks for tunnel positioning in single-bundle and double-bundle anterior cruciate ligament reconstructions. Am J Sports Med 39(4):743–752

Micro-scale Analysis of Compositional and Mechanical Properties of Dentin Using Homotopic Measurements

Anil Misra, Orestes Marangos, Ranganathan Parthasarathy, and Paulette Spencer

Abstract High resolution techniques are being increasingly applied to image or measure biomaterial properties. Data interpretations from these measurements have to be performed carefully to extract meaningful information. Two aspects become significant when such techniques are used in a complementary manner. Firstly, the measurements have to be performed using a homotopic methodology such that property correlations or 'data fusion' considers the same material volume. Secondly, appropriate mathematical models must be applied to interpret the data in terms of material properties since the high resolution technique seldom measure the properties directly. In this paper, we have described the application of scanning acoustic and scanning electron microscopy to measure the mechanical and the compositional properties of primary tooth dentin using a homotopic methodology. We have then utilized a homogenization technique in order to understand the variation in the measured elastic moduli.

A. Misra (✉)
Department of Civil, Environmental and Architectural Engineering, University of Kansas, 1530 W. 15th Street, Lawrence KS 66045, USA

Bioengineering Research Center (BERC), University of Kansas, 1530 W. 15th Street, Lawrence KS 66045, USA
e-mail: amisra@ku.edu; pspencer@ku.edu

O. Marangos • R. Parthasarathy
Bioengineering Research Center (BERC), University of Kansas, 1530 W. 15th Street, Lawrence KS 66045, USA

P. Spencer
Department of Mechanical Engineering, University of Kansas, 1530 W. 15th Street, Lawrence KS 66045, USA

Bioengineering Research Center (BERC), University of Kansas, 1530 W. 15th Street, Lawrence KS 66045, USA

D. Iacoviello and U. Andreaus (eds.), *Biomedical Imaging and Computational Modeling in Biomechanics*, Lecture Notes in Computational Vision and Biomechanics 4, DOI 10.1007/978-94-007-4270-3_7,

Keywords Dentin • Micromechanics • Acoustic microscopy • Homotopic • Density

1 Introduction

Biomaterials are highly heterogeneous with scale-dependent composition and microstructures. Advances in non-destructive high-resolution techniques have enabled location-dependent measurement of biomaterial physical, chemical and mechanical properties at micro- and nano-meter scales. When complementary methods are applied on highly heterogeneous biomaterials, it becomes necessary that the measurements are performed at the same location for their proper interpretation or cross-correlations. The authors have recently introduced the term *homotopic* (Greek *homos* = identical and *topos* = place) to describe the methodology in which a set of material properties are measured at the same location of the same sample (Marangos et al. 2009, 2011). The methodology was applied to obtain the composition and mechanical properties of a primary tooth dentin. The measurements were performed using scanning acoustic microscopy (SAM) and backscattered scanning electron microscopy (BSEM). This paper focuses upon the interpretation of these measurements using a homogenization method with the view of relating micro-scale (~100 μm) mechanical properties to the composition at the same scale.

2 Dentin Structure, Composition and Elasticity

Dentin is a hydrated composite of carbonate rich, calcium deficient apatitic nanocrystallites, type I collagen and other minor proteins that constitutes a bulk of the tooth structure (Marshall et al. 1997; Gage et al. 1989; Linde 1989; Butler 1992; Arsenault 1989; Jones and Boyde 1984). Dentin serves as a support for the overlying enamel and a cover for the pulp cavity. The hierarchical dentin structure from its molecular organization to the whole tooth scale is depicted in Fig. 1.

Typical volumetric composition of dentin is as follows: 45–50% inorganic material, 30–35% organic material and 20% fluid. The microstructure of sound dentin is unique, characterized by a system of dentinal tubules surrounded by a collar of highly mineralized peritubular dentin (Wang and Weiner 1998; Weiner et al. 1999; Spencer et al. 2007). The tubules traverse the dentin structure from the pulp cavity to the region just below the dentin-enamel junction (DEJ) or the dentin-cementum junction (CEJ). Tubule density, size and orientation vary from the pulp chamber to the DEJ. The density and size are lowest close to the DEJ and highest at the predentin surface at the junction to the pulp chamber. The porosity of dentin varies from 0 to 0.25 from the DEJ to the pulp (Manly and Deakins 1940; Sumikawa et al. 1999; Koutsi et al. 1994). In primary tooth dentin, the dentinal tubule density and size is, in general, larger than in permanent dentin

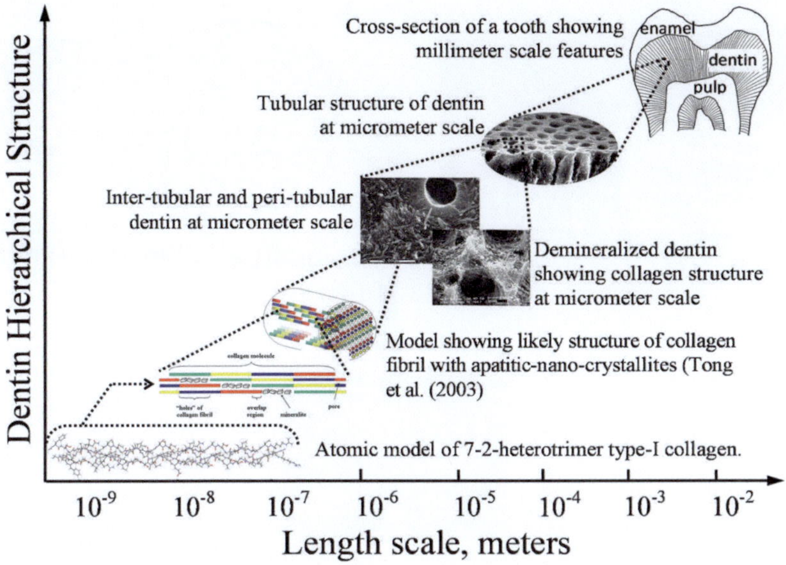

Fig. 1 A schematic depiction of dentin hierarchical structure from molecular organization to the whole tooth

(Sumikawa et al. 1999). Water in dentin is present within the dentinal tubules as pulpal fluid and within the interstitial spaces between collagen fibrils. Part of the water can be considered free and part bound. The dentin elasticity is an important feature that determines the mechanical behavior of the tooth structure. Elastic properties have been measured at different scales using a variety of methods (Kinney et al. 2003; Marangos et al. 2009). For measurements at scales on the order of 100 μm or larger, the elastic moduli has to be considered as that of a homogenized continuum depicted in Fig. 2.

Recently the authors presented measurements with SAM and BSEM of carious, caries-affected and sound primary tooth dentin using the homotopic measurement methodology (Marangos et al. 2009). As a result, location-dependent elastic moduli, density and composition of sound, carious and caries-affected primary dentin were obtained at ~100 μm spot size. In Fig. 3, we show the maps of mass density and the corresponding elastic moduli.

The homotopic measurement methodology does not rely on image matching or image processing associated with methods such as image registration, data fusion or data integration. In the image-based methods the regions of interest (ROI) are matched once the data has been acquired. Since in two images acquired in two different instruments, the objectivity of observer coordinates is not guaranteed, the pixels are not likely to represent the same material volumes. This is especially true for the ROI that have undergone rigid body motions. The key distinction of the homotopic methodology is that the measured locations are relocated in each imaging device accounting for their rigid body motion (Marangos et al. 2011). Clearly, if homotopic measurements are not made and image-based methods are

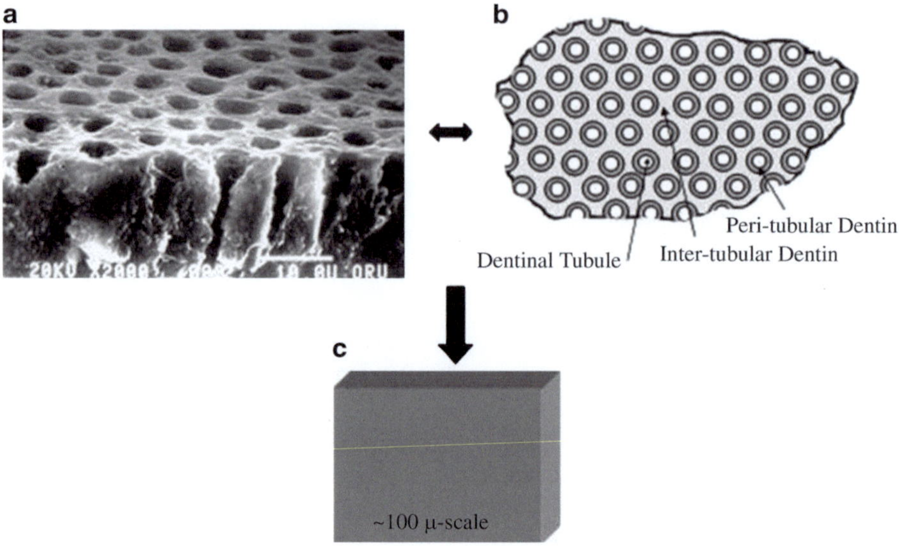

Fig. 2 (**a**) SEM micrograph of dentin microstructure showing the spatial distribution of dentin tubules; (**b**) *schematic view* of the microstructure; and (**c**) *homogenized view* for defining continuum elastic properties

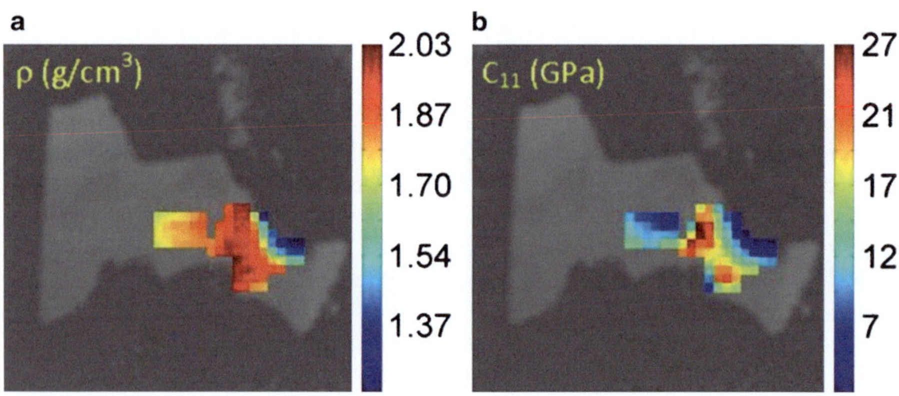

Fig. 3 Microscale distribution of (**a**) density and (**b**) elastic modulus of carious, caries-affected and sound regions of primary dentin using homotopic measurements from SAM and BSEM. *Background image* is the SAM C-scan of the tooth sample (Marangos et al. 2009)

used instead, the quantitative correlations can be misleading. For completeness, we briefly describe how the maps in Fig. 3 were constructed from the measurements.

In SAM a focused acoustic transducer is used that acts both as a transmitter and a receiver. The dentin sample is placed at the focal plane of the transducer inside a water bath. The acoustic impedance mismatch between water and the sample produces a reflected signal that is received by the transducer and analyzed to obtain the sample reflection coefficient using calibration with known materials.

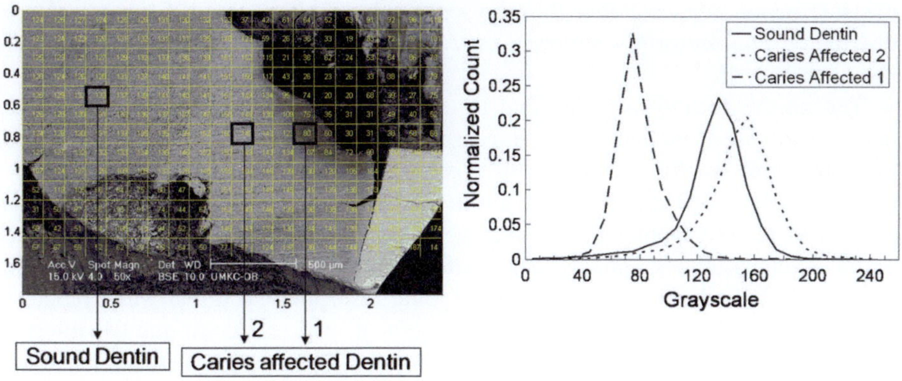

Fig. 4 (**a**) BSEM image subdivided into 120 μm square pixels corresponding to SAM pixel size, and (**b**) typical distribution of grayscales within the 120 μm pixels at three representative locations

For heterogeneous materials with soft and stiff locations, such as caries-affected dentin, the reflected signal amplitude can be influenced by the saturation effects of system electronics. The calibration has to be carefully developed to distinguish the effect of system electronics from the material response (Marangos et al. 2009). The reflection coefficient, *R*, of a sample is related to its acoustic impedance, Z (=ρc), as follows:

$$R = \frac{Z - Z_w}{Z + Z_w} \tag{1}$$

where Z_w = acoustic impedance of water (= $\rho_w c_w$ = 1.49 MRayl, at 24°C corresponding to the average temperature of the water bath), ρ denotes the material density, and $c = (C_{11}/\rho)^{1/2}$ is the longitudinal wave speed in the material with a constrained elastic modulus of C_{11}. The material elastic modulus is thus determined from the measured reflection coefficient. The material volume interrogated by the focused acoustic field determines the scale at which this elastic modulus is specified. For the maps shown in Fig. 3, a 30 MHz central frequency focused transducer with a nominal lateral resolution = ~100 μm was utilized to obtain reflected signals from a large number of locations on the specimen. The locations were spaced at 120 μm apart such that the interrogated area was slightly larger than the nominal lateral resolution of the transducer so as to minimize the overlap in the properties of adjacent locations.

Micro-scale mineral content in dentin can be determined in a relative straight forward manner by utilizing the correlation between the grayscale values of BSEM image and the weight percent of calcium constructed by imaging graphite and hydroxyapatite standards (Marangos et al. 2009). For the dentin sample shown in Fig. 3, the BSEM image was obtained at 50x magnification at a resolution of 1.7 × 1.7 μm was achieved. The regions of interest in the BSEM image were divided into square pixels of 120 × 120 μm (corresponding to SAM pixel size) as shown in Fig. 4a, to obtain lateral resolution compatibility with the SAM measurements. We note that the grayscale values within the 120 μm pixel are distributed non-uniformly although they are tightly centered on a peak value as shown in Fig. 4b.

The non-uniformity is partly due to the surface topography and partly due to material heterogeneity at the scale of measurements. For our calculations, the mineral mass fraction, m_l, for each 120 μm pixel was obtained by using the arithmetic average of the grayscale values subject to the upper and lower grayscale thresholds of 230 and 11 corresponding to the average grayscale value of the hydroxyapatite and graphite standards, respectively. The spatial mineral mass fraction variation at the 120 μm-scale is due to: (1) dentin undergoing resorption such that water replaces mineral, (2) additional water replacing mineral in carious dentin and (3) mineral displacing tubule water in caries-affected dentin. The change in mineral content results in change of water content and consequent change of density from its reference state. Considering the relative proportion of the mineral, organic component and water, the average density, ρ, for each 120 μm pixel was subsequently computed as follows:

$$\rho = \frac{\rho^{ref} m_I^{ref} + \left(\frac{\rho_I}{\rho_w}\right)\left(\rho_w \phi_{vw}^{ref} + \rho_w \phi_{bw}^{ref} + \rho_o \phi_o^{ref}\right)}{m_I + \frac{\rho_I}{\rho_w}(1 - m_I)} \tag{2}$$

assuming the following reference values for primary tooth dentin close to the pulp chamber: tubule volume fraction, ${\phi_{vw}}^{ref} = 0.20$, bound water volume fraction, ${\phi_{bw}}^{ref} = 0.039$, organic volume fraction, ${\phi_o}^{ref} = 0.343$, mineral volume fraction, ${m_I}^{ref} = 0.624$, and density $\rho^{ref} = 1.98$ g/cm^3. In addition the following density values were assumed: mineral density, $\rho_I = 3.00$ g/cm^3, and organic component density, $\rho_o = 1.40$ g/cm^3. The reflection coefficients from Eq. 1 and the densities from Eq. 2 were combined to obtain the dentin elasticity, C_{11}, shown in Fig. 3.

3 Analysis of Elasticity Measurements Using Homogenization Model

From Fig. 3, we note that there is considerable variation in the elastic modulus. Clearly, the composition has a significant effect on the elastic modulus, and previous investigators have attempted to correlate the mineral content to elastic modulus in dentin (Angker et al. 2004; Bembey et al. 2005). If we consider the relationship of the elastic modulus and the mineral volume fraction for the 89 locations displayed in Fig. 3, we observe that a single correlation cannot be used to describe the relationship between the elastic modulus and composition (Marangos et al. 2009). It is remarkable that at certain mineral volume fraction, the elastic modulus is found to vary from 5 to 25 GPa. Thus, the volume fraction alone cannot be used to uniquely determine the elastic modulus. Similar observations have

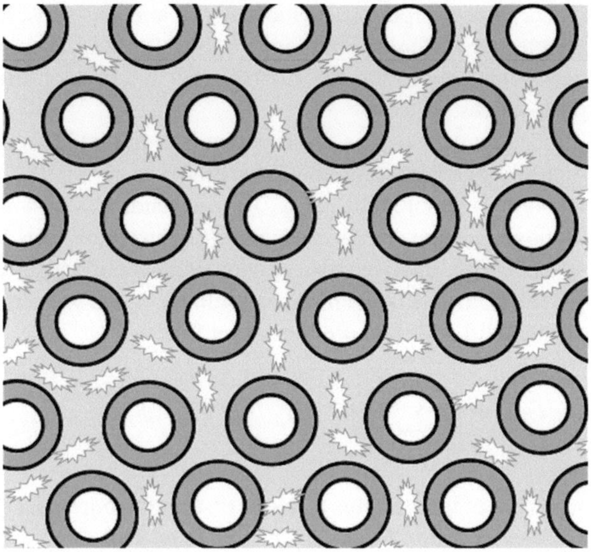

Fig. 5 Idealized microstructure of dentin showing mineral loss from dentinal matrix caused by resorption or caries

been made regarding the elastic properties of bones (Oyen et al. 2008). At the fundamental level, the interactions between collagen and nanocrystalline mineralite gives rise to the stiffness of the dentin structure. Indeed the various efforts (Hellmich et al. 2004; Nikolov and Raabe 2008) to calculate calcified tissue moduli based upon constituent phases suggest that the mechanical behavior of these tissues at various hierarchical scales may not be determined by simply considering the composition but rather additional interaction at the collagen/mineral interface must be considered.

To understand the relationship between composition and the elastic moduli, we consider a two-level homogenization scheme which can bound the data spread. Since the sample was undergoing resorption and had carious regions, part of the matrix mineral was replaced by water to create secondary porosity as shown schematically in Fig. 5. We expect this secondary pore space to be randomly distributed within the dentinal matrix with pore size on the order of the size of the nanocrystalline mineralite. At this submicron scale, the 3-component system formed by collagen/water/mineral is analyzed using rule of mixtures to obtain bounds of dentinal matrix elastic constants ${C^s}_{ijkl}$ as follows:

$$C^s_{ijkl} = \phi_I C^I_{ijkl} + \phi_o C^o_{ijkl} + \Delta\phi_w C^w_{ijkl} \qquad \text{(upper bound Voigt average)} \qquad (3)$$

$$S^s_{ijkl} = \phi_I S^I_{ijkl} + \phi_o S^o_{ijkl} + \Delta\phi_w S^w_{ijkl} \text{ and } C^s_{ijkl} = \left(S^s_{ijkl}\right)^{-1} \qquad \text{(lower bound Reuss average)} \qquad (4)$$

where superscripts on elastic tensors and the subscripts on the volume fractions denote the different components as follows: I = mineral, o = organic, and w = water, and $\Delta\phi_w$ denotes the volume fraction of the water saturated secondary pore-spaces.

The dentinal matrix elastic bounds obtained from Eqs. 3 and 4, are incorporated in a homogenized porous media model to obtain the elastic constants at ~120 μm resolution denoted by C_{ijkl}. To this end, a self-consistent approach is used to homogenize the idealized microstructure shown in Fig. 5, which is treated as a solid dentinal matrix pierced with tubular pore spaces.

The homogenized elastic modulus, C_{ijkl}, is written as weighted volume average of the solid and the pore phases as follows:

$$C_{ijkl} = \frac{1}{V} \sum_{p=s,\phi} V^p C^p_{ijmn} H^p_{mnkl} = (1-\phi)\, C^s_{ijmn} H^s_{mnkl} = C^s_{ijkl} - \phi\, C^s_{ijmn} H^\phi_{mnkl} \tag{5}$$

where ϕ is the volume fraction of tubular pore space, the pore space elasticity is identically zero ($C^\phi{}_{ijkl} = 0$), and $H^s{}_{ijkl}$ and $H^\phi{}_{ijkl}$ are concentration tensors that satisfy the identity:

$$I_{ijkl} = (1-\phi)\, H^s_{ijkl} + \phi\, H^\phi_{ijkl} \tag{6}$$

where I_{ijkl} is a fourth-rank identity tensor. The concentration tensor $H^\phi{}_{ijkl}$ can be obtained in terms of the Eshelby tensor (Mura 1987) denoted here by E_{ijkl} as follows:

$$H^\phi_{ijkl} = \left(I_{ijkl} - E_{ijkl}\right)^{-1} \tag{7}$$

The Eshelby tensor, E_{ijkl}, is given in terms of the unknown homogenized elastic modulus, C_{ijkl}, following the self-consistent approach. For further calculations we assume that the dentin behavior at the ~120 μm scale is isotropic. Such an assumption is reasonable based upon the measurements reported by Kinney et al. 2004 using resonant ultrasound technique on a ~1 mm-scale sample. These measurements show that under dry or 'drained' condition the dentin elastic modulus at the millimeter scale is nearly isotropic. In addition, we assume that our measurements are under drained conditions since, at the scale of the measurements, the dentinal tubules are obstruction free at either end, and consequently, should be freely draining. Therefore, we do not compute the Biot's parameters, although they can be obtained in a straightforward manner in the self-consistent homogenization as

$$B_{ij} = \phi\, \delta_{mn} H^\phi_{mnij}; \qquad \Gamma = \delta_{ij} \left(C^s_{ijkl}\right)^{-1} (B_{kl} - \phi\, \delta_{kl}) + \frac{\phi}{K_f} \tag{8}$$

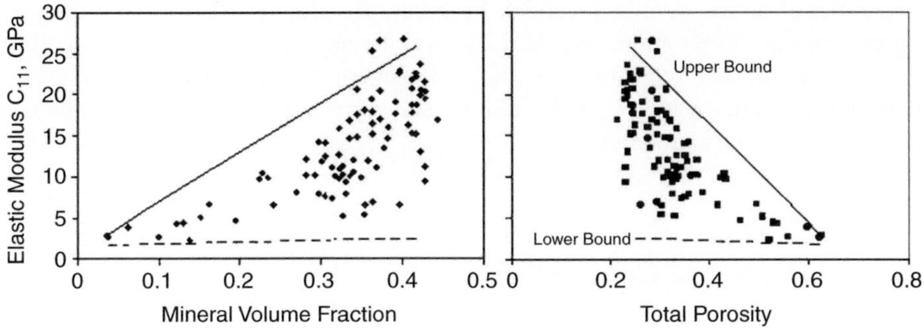

Fig. 6 Elastic modulus plotted against (**a**) the mass density, and (**b**) mineral volume fraction (Using data from Fig. 2 (Marangos et al. 2009))

where K_f is the fluid bulk modulus and the complete constitutive law is written as:

$$\sigma_{ij} = C_{ijkl}\varepsilon_{kl} - B_{ij}p; \quad \varsigma = B_{kl}\varepsilon_{kl} + \Gamma p \tag{9}$$

The fluid pressure in the tubule, p, vanishes for the drained case considered here, and our interest is confined to the elastic modulus, C_{ijkl}. For further discussion we use the Voigt notation for the elastic modulus and an equivalent notation for the concentration tensor. The two required constants for the concentration tensor of an isotropic material is given as:

$$H_{11} = \frac{C_{11}(3C_{11} - C_{12})}{(C_{11} + C_{12})(C_{11} - C_{12})}; \quad H_{12} = \frac{C_{11}(C_{11} - 3C_{12})}{(C_{11} + C_{12})(C_{11} - C_{12})} \tag{10}$$

And the homogenized elastic moduli of the isotropic media are given as:

$$C_{11} = C_{11}^s - \phi C_{11}^s H_{11} - \phi C_{12}^s H_{12}; \quad C_{12} = C_{12}^s - \phi C_{11}^s H_{12} - \phi C_{12}^s H_{11} \tag{11}$$

Combining Eqs. 10 and 11 and using the bounds of dentinal matrix elastic constants from Eqs. 3 and 4, the bounds for dentin elastic modulus can be obtained. These bounds are plotted in Fig. 6 along with the measured data as a function of the mineral volume fraction and the total porosity (given as $\Delta\phi_w+\phi$).

The tubular porosity, ϕ, was taken to be the reference value of 0.20. Furthermore, in these calculations the values of the Young's moduli and the Poisson's ratios were taken as follows: for mineral E = 100 GPa, ν = 0.23 (Gilmore and Katz 1982; Ching et al. 2009) and for organic content E = 2.4 GPa, ν = 0.28 (Katz et al. 2003). The bulk modulus of water was taken as 2.3 GPa. As seen from Fig. 6, the measured data is tightly bounded by our homogenization procedure suggesting that the material behavior is governed by different structural organization and interaction mechanism of components. Locations that are close to the bounds, the structural organization and component interaction reflects either Voigt or Reuss assumptions where the simplest analogy is that to a set of springs arranged in parallel or series. At other locations, however, the structural organization and the interactions are likely more complex. Clearly, the ability to predict elastic moduli will require further exposition of the 3-dimensional structure and interactions between the components, perhaps with increasingly higher resolution techniques and complementary modeling.

4 Summary and Conclusions

Complementary high resolution techniques are being adopted for biomaterial characterization at micro- and nano-scales. We have recently presented results from scanning acoustic microscopy (SAM) and backscattered scanning electron microscopy (BSEM) of primary tooth dentin obtained using a homotopic methodology. As a result, the elastic moduli and composition of primary tooth dentin was obtained at the ~100 μm scale. The homotopic measurement methodology is essential for accurate correlation of properties measured using complementary technique since it aims to identify the same material volume. This methodology is fundamentally different from image-based methods, such as data fusion, in which the measurements are performed first and the images matched later. To understand the relationship between measured elastic moduli and composition a mathematical model was developed. The homogenization method was used to formulate a 2-level poroelastic model that relates the composition and ~100 μm scale properties of dentin accounting for dentin microstructure. The homogenization model provides a reasonable approach to elastic moduli prediction of dentin within well defined upper and lower bounds. However, the model does not provide exact predictions for the elastic moduli and the calculations suggest that additional structural and interaction must be considered for better prediction methods.

Acknowledgments This research was partly supported by: USPHS Research Grant NIH/NIDCR DE014392 and 3R01DE014392-08S109 (PS), USPHS Major Instrumentation Grant NIH/NIDCR S10 RR16710 (PS), and NSF grant CMS-0506297 (AM). The authors gratefully acknowledge Dr. Vladimir Dusevich for his help with the backscatter electron microscopy analyses and Dr. Brenda Bohaty for her assistance with the acquisition of the primary tooth specimens. The specimens were collected under a protocol approved by the Institutional Review Board (IRB) at the University of Missouri-Kansas City.

References

Angker L, Nockolds C, Swain MV, Kilpatrick N (2004) Correlating the mechanical properties to the mineral content of carious dentine-a comparative study using an ultra-micro indentation system (UMIS) and SEM-BSE signals. Arch Oral Biol 49(5):369–378

Arsenault AL (1989) A comparative electron microscopic study of apatite crystals in collagen fibrils of rat bone, dentin and calcified turkey leg tendons. Bone Miner 6(2):165–177

Bembey AK, Oyen ML, Ko C-C, Bushby AJ, Boyde A (2005) Elastic modulus and mineral density of dentine and enamel in natural caries lesions. In: Fratzl P, Landis WJ, Wang R, Silver FH (eds) Structure and mechanical behavior of biological materials. MRS, Warrendale, pp 125–130

Butler WT (1992) Dentin extracellular matrix and dentinogenesis. Oper Dent; Suppl 5:18–23

Ching WY, Rulis P, Misra A (2009) Ab initio elastic properties and tensile strength of crystalline hydroxyapatite. Acta Biomater 5:3067–3075

Gage JP, Francis MJO, Triffitt JT (1989) Collagen and dental matrices. Wright, Boston, pp 21–24

Gilmore RS, Katz JL (1982) Elastic properties of apatites. J Mat Sci 17(4):1131–1141

Hellmich C, Ulm FJ, Dormieux L (2004) Can the diverse elastic properties of trabecular and cortical bone be attributed to only a few tissue-independent phase properties and their interactions? Biomech Model Mechanobiol 2(4):219–238

Jones SJ, Boyde A (1984) Ultrastructure of dentine and dentinogenesis. In: Linde A (ed) dentine and dentinogenesis. CRC Press, Boca Raton, pp 81–134

Katz JL, Spencer P, Nomura T, Wagh A, Wang Y (2003) Micromechanical properties of demineralized dentin with and without adhesive infiltration. J Biomed Mater Res A 66A (1):120–128

Kinney JH, Marshall SJ, Marshall GW (2003) The mechanical properties of human dentin: a critical review and re-evaluation of the dental literature. Crit Rev Oral Biol Med 14(1):13–29

Kinney JH, Gladden JR, Marshall GW, Marshall SJ, So JH, Maynard JD (2004) Resonant ultrasound spectroscopy measurements of the elastic constants in human dentin. J Biomech 37(4):437–441

Koutsi V, Noonan RG, Horner JA, Simpson MD, Matthews WG, Pashley DH (1994) The effect of dentin depth on the permeability and ultrastructure of primary molars. Pediatr Dent 16(1):29–35

Linde A (1989) Dentin matrix proteins: composition and possible functions in calcification. Anat Rec 224(2):154–166

Manly RS, Deakins ML (1940) Changes in the volume per cent of moisture, organic and inorganic material in dental caries. J Dent Res 19(2):165–170

Marangos O, Misra A, Spencer P, Bohaty B, Katz JL (2009) Physico-mechanical properties determination using microscale homotopic measurement: application to sound and caries-affected primary tooth dentin. Acta Biomater 5:1338–1348

Marangos O, Misra A, Spencer P, Katz JL (2011) Scanning acoustic microscopy investigation of frequency-dependent reflectance of acid-etched human dentin using homotopic measurements. IEEE Trans UFFC 58(3):585–595

Marshall GW, Marshall SJ, Kinney JH, Balooch M (1997) The dentin substrate: structure and properties related to bonding. J Dent 25(6):441–458

Mura T (1987) Micromechanics of defects in solids, 2nd edn. Martinus Nijhoff, Dordrecht

Nikolov S, Raabe D (2008) Hierarchical modeling of the elastic properties of bone at submicron scales: the role of extrafibrillar mineralization. Biophys J 94(11):4220–4232

Oyen ML, Ferguson VL, Bembey AK, Bushby AJ, Boyde A (2008) Composite bounds on the elastic modulus of bone. J Biomech 41(11):2585–2588

Spencer P, Wang Y, Katz JL (2007) Dentin. In: Metin Akay (ed) Wiley encyclopedia of biomedical engineering. Wiley, Hoboken

Sumikawa DA, Marshall GW, Gee L, Marshall SJ (1999) Microstructure of primary tooth dentin. Pediatr Dent 21(7):439–444

Tong W, Glimcher MJ, Katz JL, Kuhn L, Eppell SJ (2003) Size and shape of mineralites in young bovine bone measured by atomic force microscopy. Calcif Tissue Int 72:592–598

Wang R, Weiner S (1998) Human root dentin: structure anisotropy and Vickers microhardness isotropy. Connect Tissue Res 39(4):269–279

Weiner S, Veis A, Beniash E, Arad T, Dillon JW, Sabsay B, Siddiqui F (1999) Peritubular dentin formation: crystal organization and the macromolecular constituents in human teeth. J Struct Biol 126(1):27–41

Microtomography and Its Application in Oral and Implant Research

Rossella Bedini, Deborah Meleo, and Raffaella Pecci

Abstract X-ray microtomography is a miniaturized form of traditional axial computerized tomography that allows three dimensional investigations on small radiopaque objects, with an high resolution (about 5–8 μm), in a non-invasive and non-destructive way.

Compared with the conventional electronical and microscopical techniques that produce only bidimensional images, microCT is used to obtain a three dimensional analysis of a sample with no need to cut and no need of particular chemical treatments at all.

Therefore, X-ray 3D microtomography may satisfy the ideal requirements of 3D microscopy:

- Investigation of a sample without preparation or alterations;
- Production of non-invasive, non-destructive 3D images with a sufficient magnification;
- Effectiveness of measurements of internal structure's numerical features (morphology, structure and ultra-structure)

This justifies the application of this innovative technique not only in medicine and odontostomatology, but also in biomedical engineering, material science, biology, electronic, geology, archeology, petroleum and semiconductors industries.

In this section of the book will be showed different possibilities of microtomographic applications in biomedical field:

Biomaterials and Bone first, will show microtomographic appliance in structural characterization during the several phases of scaffolds' designing and manufacturing, also after their application in implantology's bone regeneration and maxilla-facial surgery;

R. Bedini (✉) • D. Meleo • R. Pecci
Technology and Health Department, Italian National Institute of Health, v.le Regina Elena 299, 00161 Rome, Italy
e-mail: rossella.bedini@iss.it

D. Iacoviello and U. Andreaus (eds.), *Biomedical Imaging and Computational Modeling in Biomechanics*, Lecture Notes in Computational Vision and Biomechanics 4,
DOI 10.1007/978-94-007-4270-3_8, © Springer Science+Business Media Dordrecht 2013

Fixture-Abutment connection characterization, in the second place, will show the results of the 3D micro-characterization analysis of fixture-abutment interface to observe no correct compliances.

Keywords Microtomography • Scaffolds • Fixture-abutment interface

1 Biomaterials and Bone

1.1 Introduction

X-ray microtomography (microCT) is an interesting application of traditional computerized tomography (CT). By means of this new technology it's possible to obtain 3D reconstructions of the internal structure of small objects without sample treatment or preparation.

The aim of this book-section is to show the possible applications of micro-CT in the analysis of biomaterials implanted in bone in order to define their morphometric properties by means of a microtomography system.

The rehabilitation of damaged skeletal segments due to pathologies or traumas is an extremely difficult objective in orthopaedic and oro-maxillo-facial surgery which may be now obtained through the great progress made by scientific research and the continuous evolution of surgical techniques that allows the use of bone substitutes, but not yet with excellent results.

The ideal graft biomaterial should be:

- biocompatible and biotolerable;
- not suitable for the proliferation of pathogenic micro-organisms;
- non-carcinogenic or teratogenic;
- non-antigenic;
- non-toxic;
- hydrophilic;
- sterile or sterilizable;
- completely bioresorbable;
- easy to manipulate;
- radiopaque;
- able to support biomechanical stress;
- economic (Jones et al. 2007; van Lenthe et al. 2007; Chiapasco and Romeo 2002).

But the most important feature of all is that it should be bioactive, i.e. able to stimulate bone regeneration through osteoinduction, osteoconduction and osteogenesis.

Osteoinduction refers to bone neoformation that occurs due to biomolecules which are responsible for the differentiation of mesenchymal cells extracted from the receiving site or from chondrogenic or osteogenic elements.

When the grafted material works as a scaffold, the osteoconduction takes place: a three-dimensional matrix, working as a framework, lead the growth of osteoblasts on defection walls penetrating into the graft internal structure (i.e. creeping substitution).

Therefore, osteogenesis represent the new osseous tissue ingrowths' promoted by osteocompetent cells in the graft, divided in osteoblasts, chondroblasts and fibroblasts due to the presence of an osteoconductive matrix, acting as a support, and signal molecules, or instead growth factors and morphogenetic proteins which can stimulate cell differentiation (Chiapasco and Romeo 2002; Cowan et al. 2007; Urist 1965; Urist et al. 1983).

In order to reduce post-operative failure due to bacteria together with the need for a second surgical site in the same patient, now autologous bone remains the gold standard because it is the only material having osteogenetic, osteoinductive and osteoconductive properties (Urist 1965; Urist et al. 1983; Chu et al. 2007; Mastrogiacomo et al. 2005; Burchardt 1983), even if the need for graft biomaterials, different from autologous bone, is increasing.

Tissue engineering technology now allows a modern alternative to the traditional approach to the reconstruction of skeletal segments with the autologous cell ex-vivo growth and their use in tissue regeneration through culture in three-dimensional biocompatible matrices (Chu 2007; Mastrogiacomo et al. 2005; Komlev et al. 2006; Hofmann et al. 2007).

It has been demonstrated that in orthopaedic and maxillofacial surgery, mesenchymal stem cells, precursors of osteoblasts, can be implanted in order to allow the new bone tissue regeneration in the skeletal damage, trauma and bone cancer treatment.

Several authors (van Lenthe et al. 2007; Cowan et al. 2007; Chu et al. 2007; Mastrogiacomo et al. 2005; Komlev et al. 2006; Hofmann et al. 2007; Papadimitropoulos et al. 2007; Cancedda et al. 2007) have been widely studied and showed that the combined use of progenitor cells, biomolecules and biomaterial scaffolds, can act as osteoconductors for bone regeneration. These devices need to have biocompatibility and biodegradability properties, as they have to allow the cellular multiplication and the matrix production, ensuring vascularisation in the same time.

The industrial production of scaffold with these characteristics is one of the most important goals in bone tissue regeneration strategy. An important aid to project and improve this kind of devices is to define the morphometric architectural characteristics of scaffolds using a 3D non-destructive and non-invasive method as a powerful investigative technique like the x-ray micro Computed Tomography.

The aim of this book section is to show, through the analysis of some biomaterials of both natural and synthetic origin, the possible application of micro-CT on the displaying and morphometric characterization of implanted grafted bone material samples.

1.2 Materials and Methods

The microCT consists of a microfocus tube which generates a cone-beam of X-rays, a rotating specimen stub on which the object to observe is put, and an electronic detector system which acquires the images. The X-ray source is a 10 W microfocus tube (tungsten), which can respectively operate at currents and voltages up to 98 μA and 100 kV. The stub can rotate and it usually use a 0.45° rotation step for complete acquisition of 180° for a total time of about 2 h.

Comparing different resolutions of CT systems, it's possible to see that Medical-CT resolution is of about 100 μm and laboratory micro-CT resolution range is from 5 to 10–20 μm, depending to sample dimensions. This system is supplied with a 1 mm-thick aluminium plate, placed in front of the scintillator, in order to use it as hardware beam hardening minimiser filter.

MicroCT system used in this work, the Skyscan 1072 instrument (SkyScan, Kartuizersweg 3B, 2550 Kontich, Belgium),allows to obtain radiographic images as well as tomographic reconstruction ones, nevertheless all data acquiring and processing are software-driven.

From the frontal projection data it is then possible to reconstruct the cross-section images of the object, using the software "Cone_rec" (ver. 2.23, Skyscan, Belgium), which is based on the cone beam algorithm (14). These projection images in TIFF format were reconstructed in slice images in the BMP format. These slices were also evaluated with "CTAnalyser" software (ver. 1.11, Skyscan, Belgium) in order to obtain the 3D structure and morphometric parameters.

To be processed in this way, images have to be binarized. The process of "binarization" needs to choose a threshold value and has as a result an image composed only of black pixels (bone) and white pixels (non-bone).

After binarization, "CTAnalyser" software allows to calculate morphometric parameters and create the three-dimensional image of the internal structure of the analyzed sample from images' slices in BMP format.

For each biomaterial sample was selected an internal Region Of Interest (ROI) and histomorphometric parameters for the corresponding Volume Of Interest (VOI) were calculated. The parameters analyzed, also called "primary morphometric indices", were:

- Tissue Volume, TV, (mm^3), total volume of the volume-of-interest (VOI). The 3D volume measurement is based on the marching cubes volume model of the VOI.
- Bone Volume, BV, (mm^3), total volume of binarized objects within the VOI. The 3D volume measurement is based on the marching cubes volume model of the binarized objects within the VOI.
- Tissue Surface, TS (mm^2), the surface area of the volume of interest, measured in 3D (Marching cubes method).
- Bone Surface, BS, (mm^2), the surface area of all the solid objects within the VOI, measured in 3D (Marching cubes method).
- Bone Volume Fraction, BV/TV, (%), the proportion of the VOI occupied by binarized solid objects.
- Specific Bone Surface, BS/BV, (1/mm), the ratio of solid surface to volume measured in 3D within the VOI. Surface to volume ratio or "specific surface" is a useful basic parameter for characterizing the thickness and complexity of structures.

– Bone Surface Density, BS/TV, (1/mm), the ratio of surface area to total volume measured as described above in 3D, within the VOI, (Chiapasco and Romeo 2002; Cowan et al. 2007; Urist 1965; Mastrogiacomo et al. 2005; Perilli et al. 2007; Gielkens et al. 2008; Haïat et al. 2007; Jones et al. 2007; Kachelrieb 2008; Kamburoglu et al. 2008; Kerckhofs et al. 2008; Lin-Gibson et al. 2007; Maréchal et al. 2005; Papadimitropoulos et al. 2007; Parkinson et al. 2008; Stauber and Müller 2008; van Lenthe et al. 2007; Willems et al. 2007).

There are also "secondary indices" for the quantification of bone architecture. The model most widely used is the plate model, which assumes that all the trabecular bone is organized in infinite plates, with a certain thickness (Tb.Th), separation (Tb.Sp) and number per unit length (Tb.N) (Waarsing et al. 2004).

The "secondary indices" are:

- Trabecular Thickness,Tb.Th, (mm), that represents the thickness of the trabecular bone

$$\mathrm{Tb.Th} = 2\mathrm{BV}/\mathrm{BS}$$

- Trabecular Number,Tb.N, (1/mm), that represents the trabecular density and it is the number of plates traversed by a line of unit length perpendicular to the plates.

$$\mathrm{Tb.N.} = {}^{1}\!/\!_{2}\mathrm{BS}/\mathrm{TV}$$

- Trabecular Separation,Tb.Sp, (mm), that represents the distance between edges of the bone trabeculae (Chiapasco and Romeo 2002; Cowan et al. 2007; Urist 1965; Mastrogiacomo et al. 2005; Jones et al. 2007; Kachelrieb 2008; Kerckhofs et al. 2008; Lin-Gibson et al. 2007; Papadimitropoulos et al. 2007; Stauber and Müller 2008; van Lenthe et al. 2007).

$$\mathrm{Tb.Sp} = 1/\mathrm{Tb}(\mathrm{N} - \mathrm{Tb.Th})$$

By means of "3D Creator.exe" software it's possible to obtain the 3D inside structure of a biomaterial sample.

This software includes all the functions that are necessary to work on obtained 3D images, i.e. allows to process them with complex algorithm to obtain a better display and understanding of observed ultrastructures. (Chiapasco and Romeo 2002; Guldberg et al. 2008; Urist 1965)

In this study three commercialised osteoconductor biomaterials of different origins have been selected and named according to the following list:

1. Biomaterial 1 – Bovine hydroxyapatite, (Bio-Oss, Geistlich Biomaterials, Italy);
2. Biomaterial 2 – Bioceramic hydroxyapatite, (ENGIpore, Fin-Ceramica Faenza S.p.A, Italy);
3. Biomaterial 3 – Poli Acid (D, L – lactic – co-glicolic), (SINTBONE, GhimasS.p. A, Italy)

Bovine hydroxyapatite (biomaterial 1) is a bone substitute material for natural bone regeneration. This medical device is obtained from the mineral portion of bovine bone and is used by dental and oral surgeons in order to build up a new bone. This material is used in implantology (teeth replacement by means of implants), periodontology (for diseases affecting the tissue supporting the teeth) and for large bone defects in oral and maxillofacial surgery. Bio-Oss® supports the body's natural healing processes because of its close resemblance to human tissue. More than 700 scientific studies, in fact, demonstrated a long-term success with this material, as the product has been used in bone regeneration for 24 years.

Bioceramic hydroxyapatite (biomaterial 2) is an innovative synthetic porous hydroxyapatite biomaterial with a trabecular structure similar to natural bone. It is a ready available bone graft, manufactured to a highly controlled specification and free of disease transmission risk. Different shapes have been designed to provide surgeons with a full range of products to meet their bone grafting needs. Its particular structure gives this material a porosity of almost 90%, and this guarantees an easy access to cells, biological fluids, and signaling molecules throughout the bone substitute. Despite the highly porous structure, this material is able to resist compression forces like natural cancellous bone. Once applied in situ, the material rapidly absorbs all bio-active proteins, growth factors and bone precursor cells contained in the physiological fluids. This is the starting point of the biological cascade leading to effective bone regeneration.

Poli Acid (D, L – lactic – co-glicolic) synthetic osteoconductive biomaterial 3 is a synthetic osteoconductor biomaterial used in dental and maxilla-facial surgery, obtained from sintering process of co-polymer micro-particles (PLA/PGA). This biomaterial behaves like a rigid micro-porous texture keeping space among granules, making channels ready for vascularisation and tissue penetration. As a result of the presence of a large amount of channels in this biomaterial, blood, plasma and serum can easily get through it, causing the immediate formation of a stable blood clot making both inside and outside the biomaterial.

Each biomaterial as shaped block, in preliminary phase, has been subjected to microtomographic analysis to evaluate the most important morphometric characteristics, before clinical use. Bovine hydroxyapatite was cancellous bone block of ($13.3 \times 9 \times 6.5$) mm dimensions, Bioceramic hydroxyapatite was block of ($10 \times 5 \times 5$) mm dimensions, Poli Acid was block of ($16 \times 5.5 \times 6$) mm dimensions.

Then, the same biomaterials shaped in granular version have been glued in an ideal human bone defect.

Three healthy patients, aged 18–36, needing for a surgical dental extraction of lower third molars, have been selected.

Each patient was informed and has signed an informed consent form according to the declaration of Helsinki that requires the patient adhesion to the clinical experimental study.

In the post-extractive side of each patient has been implanted one of the biomaterials (test sample) used in this study, while the opposite side has been treated only with blood clot (control sample).

Table 1 Microtomographic acquisition parameters used for bone sample analysis

Acquisition parameters of microtomographic instrument	
Magnification (cross-section pixel size)	95X (3.1 μm)
Rotation angle	180 ± 1°
Rotation step	0.45° ± 0.01°
Power source	100 KV/98 microA ± 10%
Filter thickness (Al)	(1 ± 0.01) mm

After 4 months from surgery, test and control sample follow-up have been performed to extract cylindrical bone sample by means of stainless-steel trephine bone cutter with 2 mm internal diameter.

Each extracted bone sample was fixed in 10% formalin and before microtomographic analysis has been carefully washed with physiological solution and fixed on suitable stub by plasticine.

After microtomographic analysis, bone samples have been again inserted in sterile physiological solution for about 10 min for rehydration, and then in 10% formalin, to store them for histological investigations. Each bone sample was embedded in paraffin and 7 μm sections were prepared and stained with hematoxylin and eosin.

Each histological obtained image, for test and control bone samples, will allows to display new bone formation and scaffolding residues.

For each test and control sample the morphometric parameters have been calculated and 2D – 3D dimensional images were processed by means of the same test protocol used for microtomographic analysis of biomaterials shaped as block, before the implant phase.

The biomaterial blocks have been acquired and processed with different resolution because of different size that are commercialized, according to the following values:

- Biomaterial 1 block at 25 X magnification, corresponding to 11.72 μm pixel size;
- Biomaterial 2 block at 40 X magnification, corresponding to 7.32 μm pixel size;
- Biomaterial 3 block at 25 X magnification, corresponding to 11.72 μm pixel size.

For all control and test samples, the same acquisition parameters have been chosen (as showed in Table 1) to allow comparing qualitative and quantitative results.

To obtain a three-dimensional structure's specimen, the range of values that corresponds to image grayscale should vary according to material density. In fact, in radiology, 255 grey's levels correspond to different X-ray absorption coefficient of sample structures. It is known that absorption coefficient depends on the density of crossed material, especially the denser material that absorbs more rays and can be identified with white color, on the other hand X-rays can penetrate the less dense material, arriving on the CCD camera and be identified with black color. In this study images present inverted colours, which means that white color is used for less dense material and black for the denser.

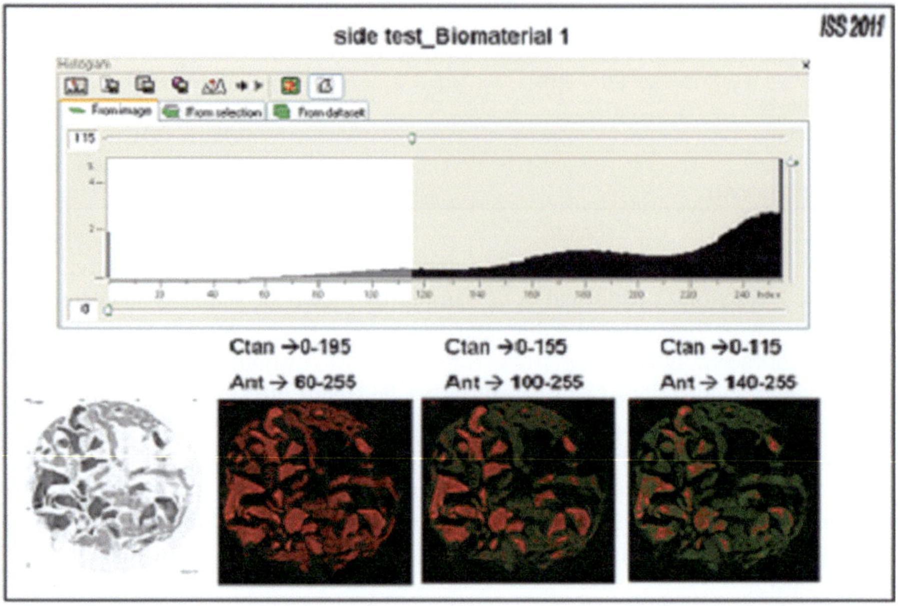

Fig. 1 On the *top* image there is the histogram for the choice of the threshold, while in the *bottom* we can see, on the *left*, one b/w slice of the biomaterial and, on the *right*, three slices with three different threshold ranges to discriminate structures with different density

The observation of images' samples showed that in all the samples three main shades of gray can be identified which can correspond to three different bone densities. Therefore, three threshold ranges were chosen (0–195, 0–155, 0–115), as showed in Fig. 1. For each range, a different color was selected to discriminate structures with different density in 3D image of the sample: yellow represents the less radiopaque component which corresponds to bone, orange represents areas that reached a first stage of calcification, and, at last, red represents areas that reached a higher stage of calcification (Fig. 2).

1.3 Results

By means of software "3D Creator.exe" it's possible to obtain the 3D structure (Fig. 3) of a biomaterial sample and also 3D slices of cut surface in any direction (Fig. 4).

For each control and test bone sample, by means of microtomographic analysis, Rx-Images and horizontal slices have been obtained, cutting at different level of sample body to display the structure of sample's inner section (Figs. 5, 6, and 7).

The morphometric parameters showed in Table 2, have been obtained as results with "CTAnalyser" Skyscan software processing for each control and test bone sample

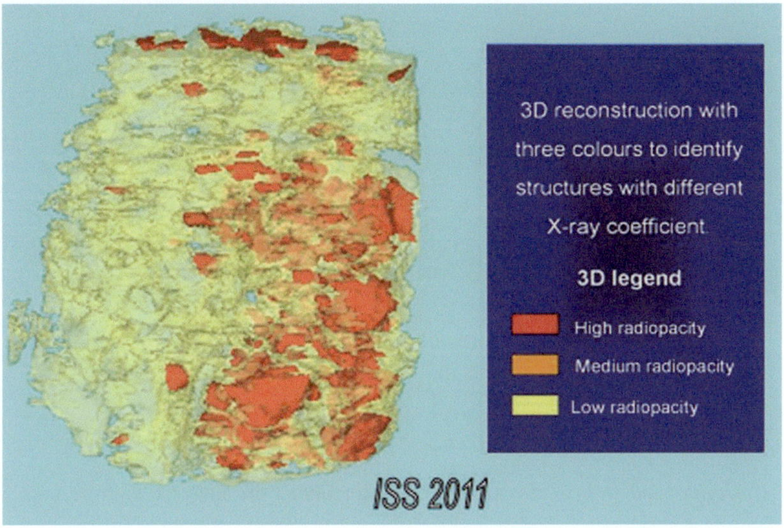

Fig. 2 Example of 3D-structure of sample grafted with biomaterial 2 and a legend of colors to explain different radiopacity of the sample

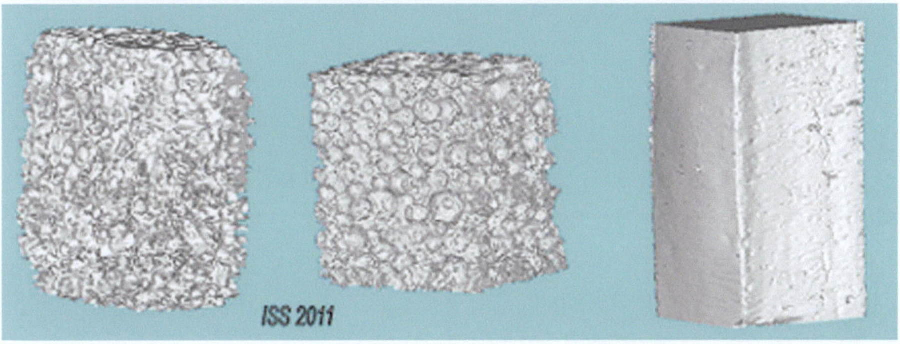

Fig. 3 3D reconstruction of the analyzed biomaterials: from *left* to *right* we can see biomaterial 1, biomaterial 2 and biomaterial 3

Test samples showed, for each biomaterial group, all BV/TV values lower than control samples ones, while BS/BV showed higher values than control samples.

Porosity values seem to have increased in test samples of Bovine hydroxyapatite and Bioceramic hydroxyapatite.

Light differences between test and control samples, for Tb.Th. parameters, have been obtained, while Tb.Sp. values for test and control samples are similar only for Poli Acid (D, L – lactic – co-glicolic), while Bovine hydroxyapatite and Bioceramic hydroxyapatite showed test bone sample values higher than control bone samples.

3D image reconstruction has been obtained by means of 3D_creator software, identifying with different colours, according to a specific code related to sample

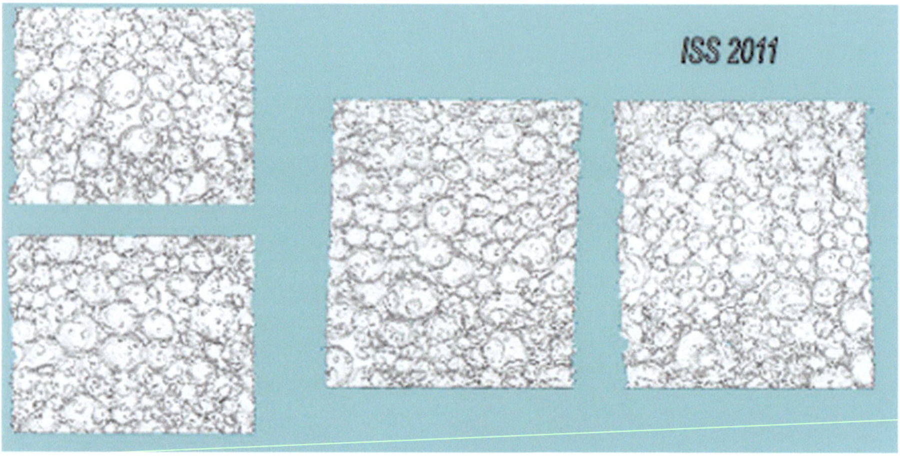

Fig. 4 *Horizontal* and *vertical* slices of 3D reconstruction of the analyzed biomaterial 2

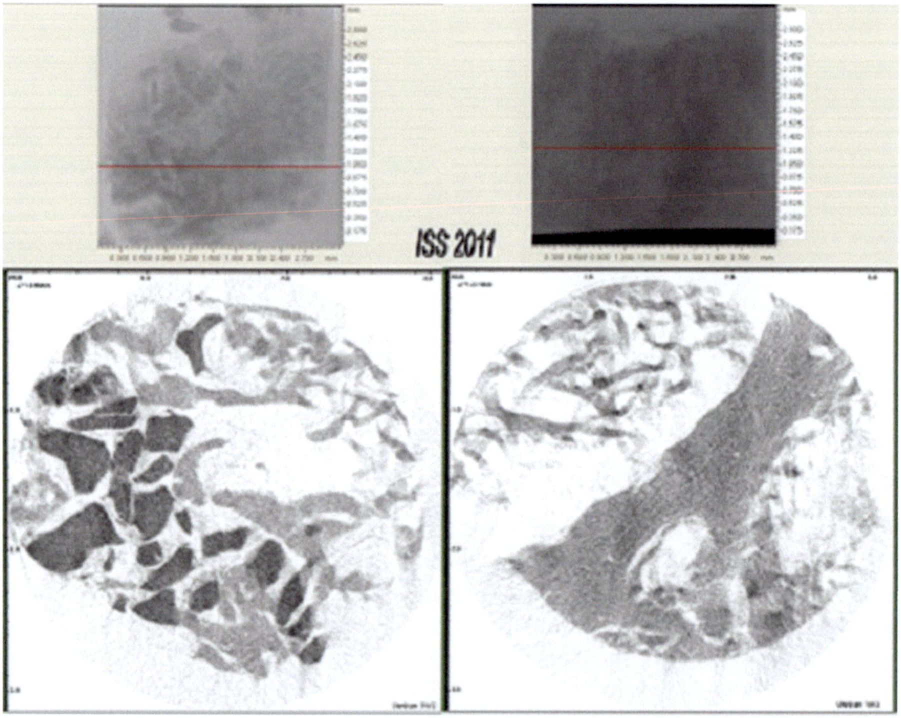

Fig. 5 Test (*left*) and control (*right*) sample slices with biomaterial 1

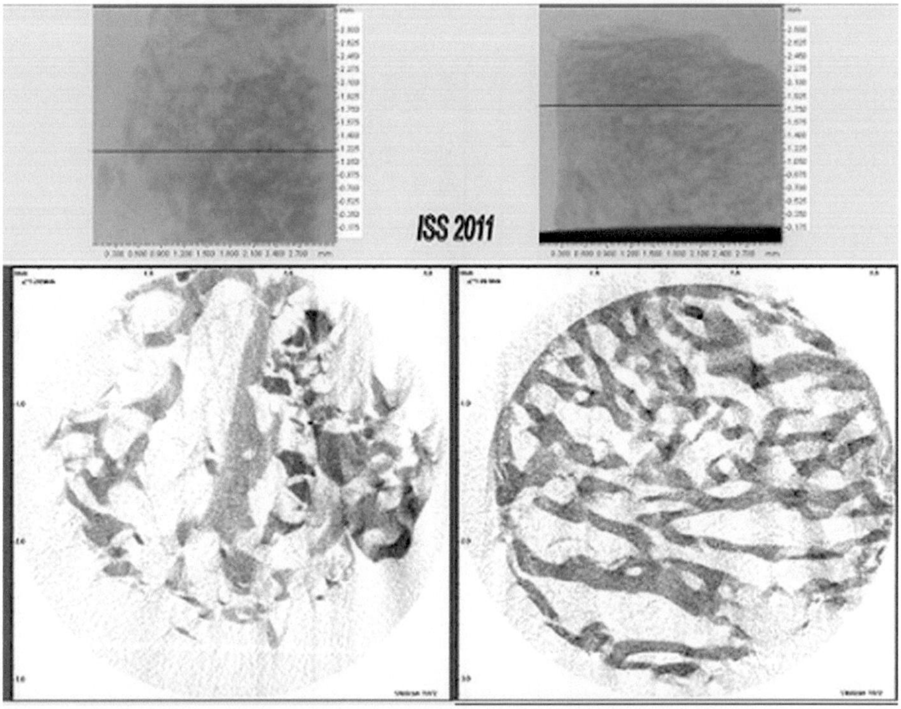

Fig. 6 Test with biomaterial 2 (*left*) and control (*right*) sample slices

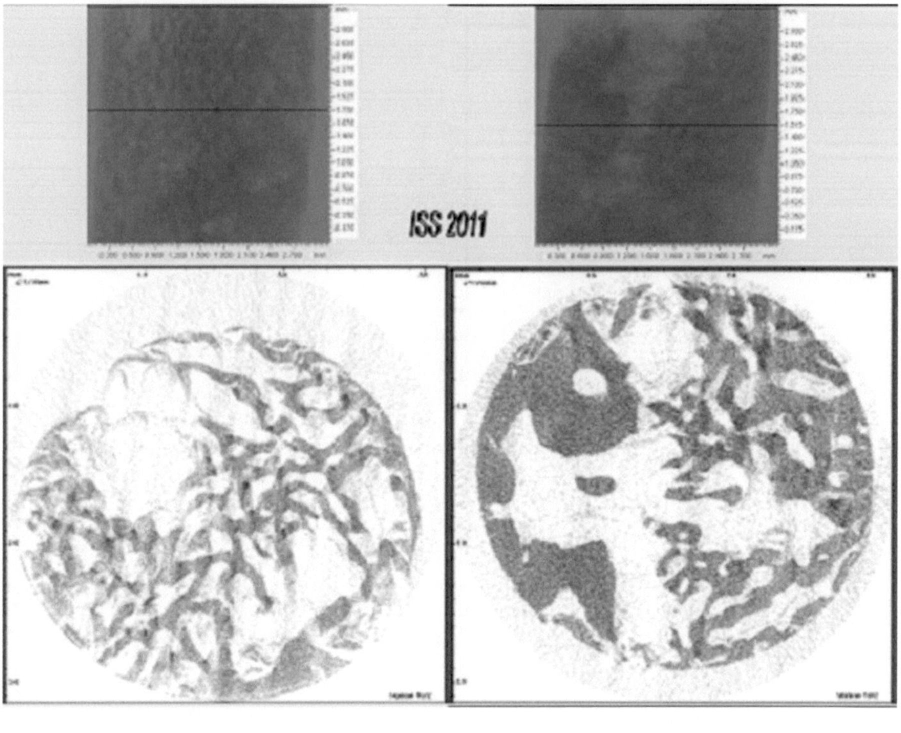

Fig. 7 Test with biomaterial 3 (*left*) and control (*right*) sample slices

Table 2 Hystomorphometric parameters obtained for control and test bone samples

Hystomorphometric values of control and test bone samples						
	Biomaterial 1		Biomaterial 2		Biomaterial 3	
Parameters	Control	Test	Control	Test	Control	Test
BV/TV (%)	28.17 ± 0.13	19.80 ± 0.05	39.25 ± 0.12	23.35 ± 0.02	39.00 ± 0.04	30.99 ± 0.08
BS/BV (mm^{-1})	90.22 ± 0.05	95.59 ± 0.05	81.06 ± 0.03	84.13 ± 0.03	26.45 ± 0.03	44.15 ± 0.01
Porosity (%)	72.20 ± 0.05	80.20 ± 0.03	60.75 ± 0.03	76.65 ± 0.03	61.00 ± 0.07	69.01 ± 0.01
Tb.Th. (μm)	0.06 ± 0.01	0.08 ± 0.01	0.06 ± 0.01	0.07 ± 0.01	0.17 ± 0.02	0.13 ± 0.02
Tb.Sp. (mm^{-1})	0.22 ± 0.02	0.37 ± 0.03	0.09 ± 0.01	0.17 ± 0.03	0.27 ± 0.03	0.27 ± 0.01

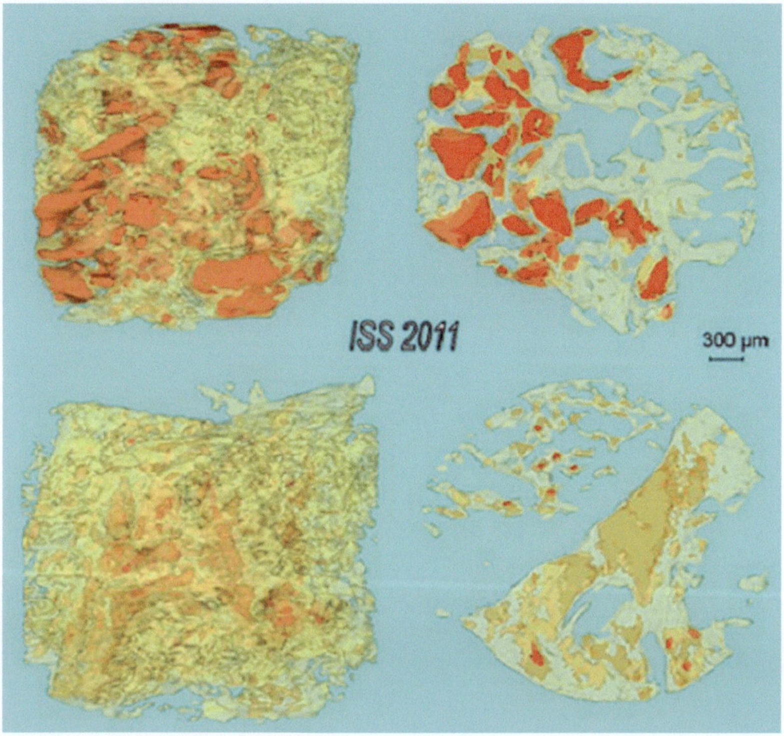

Fig. 8 *Top* image shows 3D reconstruction color images of biomaterial 1 test sample (all sample on *left* and a slice of the same sample on the *right*) while *bottom* shows 3D reconstruction color images of biomaterial 1 control sample (all sample on *left* and a slice of the same sample on the *right*)

density value, which one have a better display of used biomaterial and bone. In the images that have been showed below there is, on the left, the 3D total sample reconstruction while, on the right, there is only a slice of the same sample (Figs. 8, 9, and 10).

As mentioned earlier, yellow colour area represents the less radiopaque component which corresponds to bone, orange colour represents area that reached a first stage of calcification, and, at last, red colour represents areas that reached a higher stage of calcification.

Comparing the images of test and control sample for each biomaterial used in this study, it's possible to observe that more red areas are displayed on biomaterial 1 test sample slice image, according microtomographic analysis developed in this study.

All control sample slices show few red areas but very large yellow and orange areas.

By means of histological investigation the following images (Figs. 11, 12, and 13) have been obtained, where in control samples can be possible to observe an interesting quantity of neo-formed bone. In test samples neo-formed bone seems to be lower and it has been possible to observe also an amount of fibrous connective tissue together scaffolding residues.

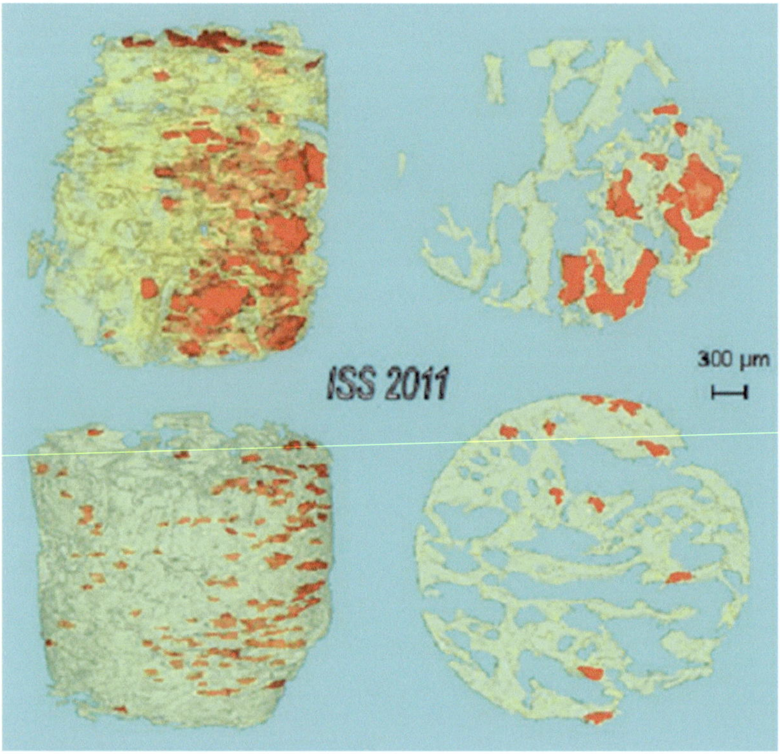

Fig. 9 *Top* image shows 3D reconstruction color images of biomaterial 2 test sample (all sample on *left* and a slice of the same sample on the *right*) while *bottom* shows 3D reconstruction color images of biomaterial 2 control sample (all sample on *left* and a slice of the same sample on the *right*)

1.4 Discussion

X-ray computed tomography is an innovative investigation technique used to study histomorphometric characteristics of a bone substitute in a non-destructive, non-invasive way. MicroCT analysis allows to analyze a sample by means of slices taken at different levels and directions using a 3D reconstruction dedicated software (Chiapasco and Romeo 2002; Cowan et al. 2007; Urist 1965).

The aim of this study is to find the most suitable bone substitute biomaterial for a specific clinical and surgical application.

The main features that the ideal scaffold should have, are: high porosity, a large surface for cellular adhesion and large pores to enable the penetration of neoformed vascular structures (Jones et al. 2007; van Lenthe et al. 2007; Chiapasco and Romeo 2002; Cowan et al. 2007; Urist 1965; Urist et al. 1983; Chu et al. 2007;

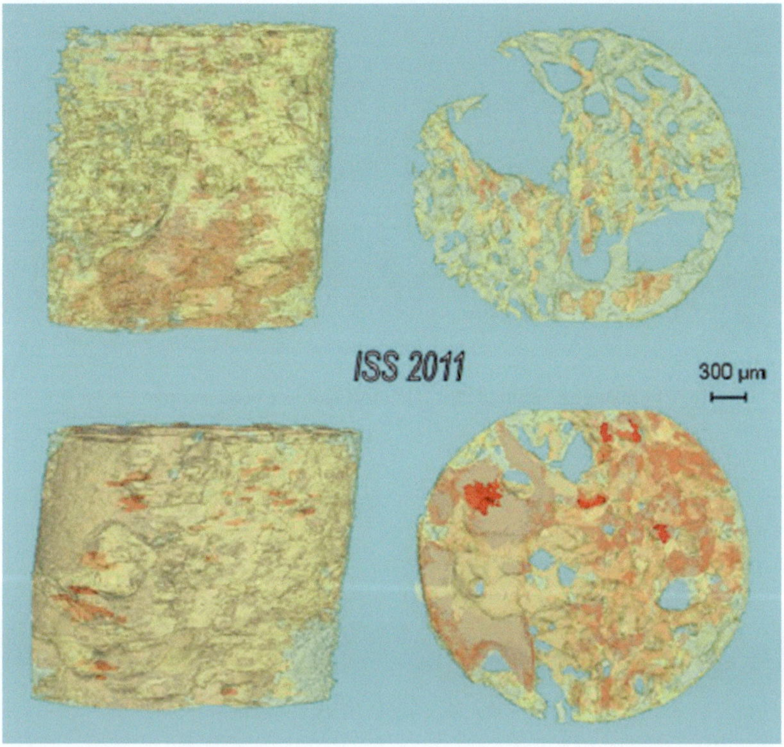

Fig. 10 *Top* image shows 3D reconstruction color images of biomaterial 3 test sample (all sample on *left* and a slice of the same sample on the *right*) while *bottom* shows 3D reconstruction color images of biomaterial 3 control sample (all sample on *left* and a slice of the same sample on the *right*)

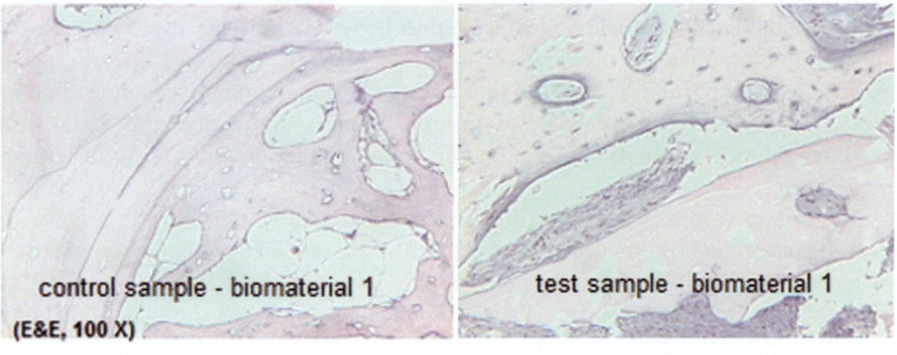

Fig. 11 Histological images of biomaterial 1 control and test bone sample

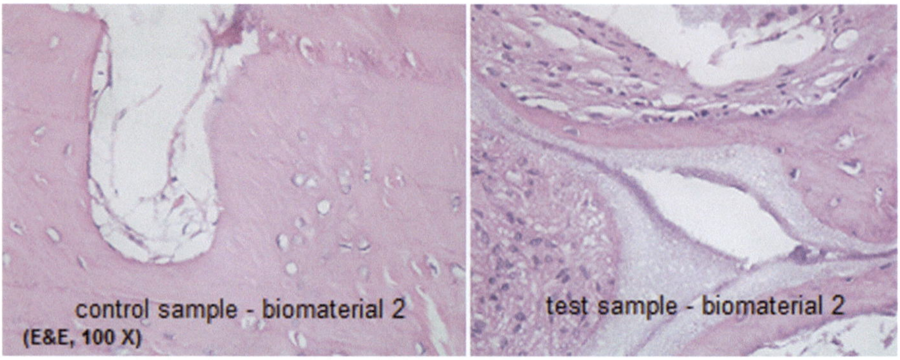

Fig. 12 Histological images of biomaterial 2 control and test bone sample

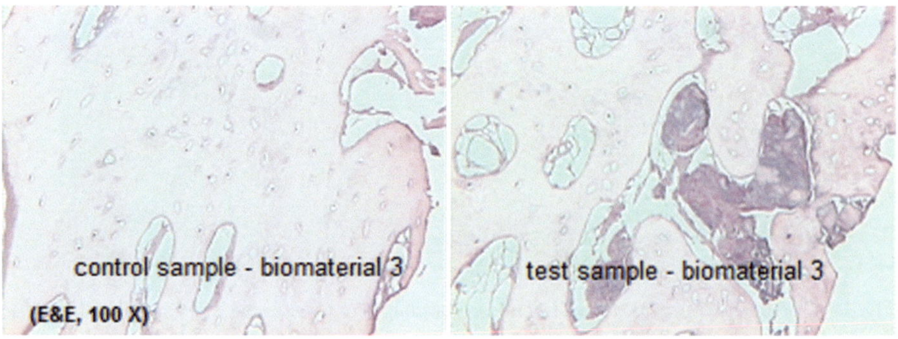

Fig. 13 Histological images of biomaterial 3 control and test bone sample

Mastrogiacomo et al. 2005; Burchardt 1983; Komlev et al. 2006; Hofmann et al. 2007).

The obtained results showed that equine bone tissue is the more suitable in substituting a human one because of its porosity and the values of Specific Bone Surface and Trabecular Separation that allow neoformed bone tissue migration in the scaffold. Other values as Bone Volume Density and Trabecular Thickness show mechanical properties of the equine bone comparable with those of human bone tissue.

The obtained results for the synthetic biomaterial show that have less porosity than the others and the small value of trabecular separation ensure the colonization of osteogenetic cells and neoformed vessels. Furthermore, lower porosity and a high value of Bone Volume Density and Trabecular Thickness show that this bone substitute biomaterial has a great mechanical resistance.

From the analysis of 3D reconstructions, interconnectivity can be observed in the negative 3D reconstructions images showing the presence of communicating spaces and connections among the pores, which are very important for cell and fluid migration within the scaffold of biomaterial. The synthetic material shows lower interconnections into the empty spaces of all the volume.

The histological images confirm what was stated above and to complete what methodology is not able to emphasize. Despite the undoubted advantages of microtomographic technique compared to the traditional optical microscopy, only histological analysis has been able to confirm the presence of biomaterial scaffolding residues in biomaterial 3 test bone sample, probably due to lower radiopacity and fine powder formulation.

At the end it's interesting to highlight that only histological images can display the presence of different no-calcified tissues, like fibrous connective tissue, vascular structures, adipose tissue and gaps that during 3D microtomographic reconstruction it's impossible to colour.

1.5 Conclusions

This study wants to be a contribution to the application of micro-CT in biomedical engineering and surgical treatments. In fact the use of micro-CT technology is becoming widespread in Europe (Perilli et al. 2007; Feldkamp et al. 1984; Feldkamp et al. 1989; Freed et al. 2006).

This innovative method produces 3D images by means of a non-destructive investigation of small objects' microstructure with no need of particular treatments (Chiapasco and Romeo 2002; Cowan et al. 2007; Urist 1965).

The main objective of this study is to provide surgeons with the most fitted bone substitute to use as a scaffold in specific clinical applications.

Nowadays, trying to avoid the use of homologous grafts, considered as a gold standard in bone regeneration, and avoid surgery in different sites in a same patient with infection risk, many researchers are developing scaffolds doped with osteoinductive factors to obtain best osteoconductive characteristics (Jones et al. 2007; van Lenthe et al. 2007; Chiapasco and Romeo 2002; Cowan et al. 2007; Urist 1965; Urist et al. 1983; Chu et al. 2007; Mastrogiacomo et al. 2005; Burchardt 1983; Komlev et al. 2006; Hofmann et al. 2007). For this purpose, MicroCT analysis could be a powerful aid to engineering and project best scaffold materials.

Therefore, this technology may be a contribution to:

- bone tissue morphometric characterization grafted with a bone substitute;
- biomaterials industry, in the production of ideal structured scaffolds;
- the study bone tissue and scaffolds' behaviour subjected to mechanical, chemical and thermal stress.
- tissue engineering 3D design of biocompatible matrix allowing ex-vivo cultivation of autologous cells used in tissue regeneration, and for the new formed bone tissue characterization, both in-vitro and ex-vivo, as a result from progenitor cells, biomolecules and biomaterials osteoconductors combination.

2 Fixture-Abutment Connection Characterization

2.1 Introduction

Nowadays dental implantology has reached high fidelity and predictability levels that were just unthinkable a few years back, allowing clinics to perform very complex rehabilitations.

In the last 30 years, the improvement of surgical procedures together with a huge technological progress contributed to the growth of successful rehabilitations on implants over 90%, giving to dentists and maxillofacial surgeons a therapeutic milestone, even if there are still a lot of problems to solve and the predictability of a rehabilitative treatment about implants requires a dynamic balance between biological and mechanical factors.

The final objective of an implant-prosthetic therapy is to obtain an esthetical valid prosthesis, but most of all a functional one, trying to avoid an eventual failure of an implant system's component (Albrektsson 1988; Oh et al. 2002; Quirynen et al. 2002).

The causes of an implant failure can be divided into two big categories: biological and mechanical. The first are mainly due to the peri-implantitis process that is a destructive inflammatory process affecting the soft and hard tissues surrounding dental implants, while the second ones are due to a component of a whole implant-prosthetic rehabilitation system. Mechanical causes can be identified in presence of fixture or abutment's fracture, screw loosening or unscrewing and prosthetic ceramic or metal superstructure fracture (Albrektsson 1988; Oh et al. 2002).

A no-correct implant surgery methodology, the presence of parafunction like bruxism, clenching and rocking of teeth or other problems related to projecting and producing implant and prosthetic components can cause a mechanical failure, occurring together or each one at a time (Albrektsson 1988; Oh et al. 2002; Quirynen et al. 2002).

The development of dental implant manufacturing allows clinics to choose on a large scale of implant systems and techniques, but in spite of a great choice of possibilities, these systems use only 3 different connections: screwed, cemented and conometric type. The screwed connection is the most famous and applied, i.e. the traditional external hexagon according to Swedish school.

In literature, in almost every study about biomechanical problems, there is this kind of connection, as most used and more problematic than the others because can be subjected to unscrewing of connecting screw, screw or implant's neck fracture (Ricomini Filho et al. 2010; Gratton et al. 2001; Mangano et al. 2009; Dibart et al. 2005; Weng et al. 2008; Persson et al. 1996; Binon 1996a, b, 2000; Brunski 1995).

To avoid all these problems, new types of connections have been projected. The anti-rotational mechanism of fixtures has been modified through the years, reaching a new shape far from traditional ones (hexagon, octagon), leading to a pure conometric type or combining this one with other traditional structures (Dibart et al. 2005; Binon 1996a, b, 2000; Brunski 1995; Bianchi et al. 2002).

The maintaining and the stability of connections with tighten screw can be put at risk by the application of unscrewing strengths occurring between implant and abutment, in fact according to bolted joint mechanics principles the probability of screws' unscrewing decrease by reducing the implant-abutment gap (Coelho et al. 2008; McGlumpy et al. 1998; Bickford 1981; Jorneus et al. 1992; Patterson and Johns 1992).

A clear observation of implant-abutment micro gaps is possible by means of various instruments that, however, can have limits in displaying them. In vivo observations are made by means of traditional methods of intra and extra-oral radiographic investigations and by computed microtomography in order to evaluate the stability or a failure of an implant (Yip et al. 2004; Bragger 1998; Rack et al. 2010; Zipprich et al. 2007). On the other hand, there are a few studies discussing in vitro observation possibilities and they're usually applied to geometrical connections with butt-joint connection: micro-radiography, SEM, optical microscopy, laser scanning microscopy or theoretical approaches using mathematical models with finite elements (Rack et al. 2010; Zipprich et al. 2007; Tsuge et al. 2008; Coelho et al. 2007; Hecker et al. 2006).

Only one study today shows the possibility to directly observe the in vitro micro gap fixture abutment of conic connections by means of hard X-ray synchrotron radiation, because in the past many authors believed that a total seal was assured by this connection geometry (Rack et al. 2010). As a matter of fact, new leaking tests demonstrated the impossibility of ensure a total seal (Rack et al. 2010; Zipprich et al. 2007; Tsuge et al. 2008).

The main purpose of this study is to show the several possibilities of X-ray microtomography applications and 3D inner displaying of implant-abutment interfaces in vitro and static conditions, analysing different kind of fixture-abutment system with different connections. This new method of investigation propose itself as a valid aid in designing, developing and manufacturing implant system with ideal features.

2.2 Materials and Methods

For the evaluation of abutment-fixture interface, and for the measurement of the contact surface, 4 *in-vitro* implant systems with different connection features have been analyzed by microtomographic technique.

The selected acquired dental implant systems were classified according to the following list:

1. A dental implant system (fixture diameter 4 mm) with an internal exagonal and platform switching connection between implant and abutment (NanoTite Certain PREVAIL, Biomet 3i, USA).

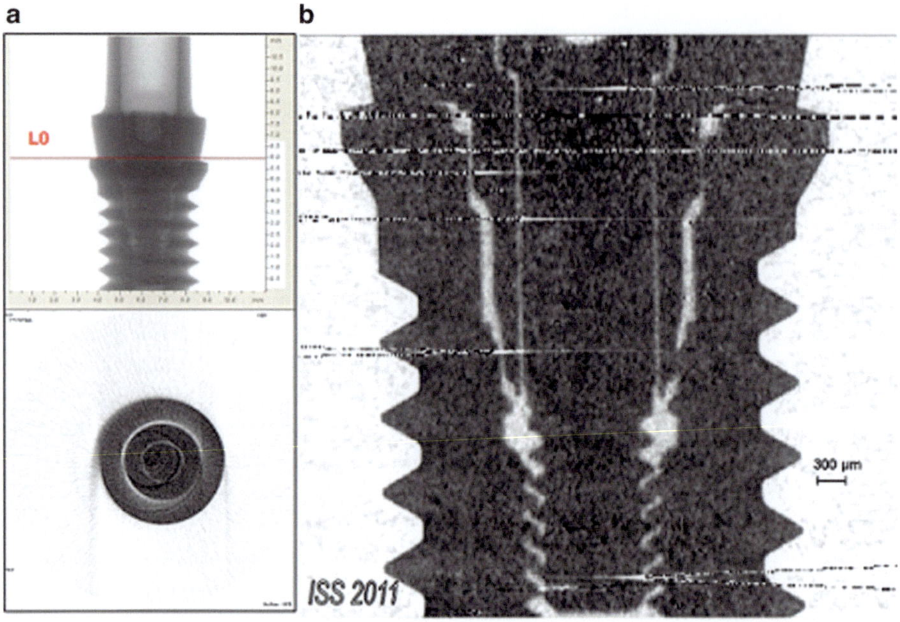

Fig. 14 Microtomographic horizontal (**a**) and vertical (**b**) section of Sample 1 with its Rx image (left corner) and indication of the L0 level

2. A dental implant system (fixture diameter 4.3 mm) with internal tri-channel connection between implant and abutment (NobelReplaceRP with Snappy Abutment, Nobel Biocare, Switzerland)
3. A dental implant system (fixture diameter 4.3 mm) with a internal conical connection – with built-in platform shifting -between fixture and abutment (Nobel ActiveRP, Nobel Biocare, Switzerland)
4. A dental implant system (fixture diameter 4 mm) with a internal conical connection between fixture and abutment and micro thread around neck of implant (OsseoSpeed, Astra Tech, Sweden).

Samples 1 and 2 present a horizontal and flat connection between implant platform and the base of the abutment, while samples 3 and 4 show a conical contact surface inside the implant body (Figs. 14, 15, 16, and 17). Furthermore, sample n.1. shows an hexagonal shaped anti-rotational system inside the implant body, while sample n.2 shows an internal tri-channel connection. In sample n.3 it's possible to observe a conic surface with tightened screw, while in sample n.4 the screw is integrated in the apical portion of the conical part of the abutment.

Each sample has been subjected to 5 consecutive acquisitions by means of X-ray microtomography with Skyscan 1072 instrument (SkyScan, Kartuizersweg 3B, 2550 Kontich, Belgium) to obtain the measurements of contact area mean values from the 4 different connections analyzed. This innovative investigating technology allowed to evaluate in a non-destructive, non-invasive and three-dimensional way

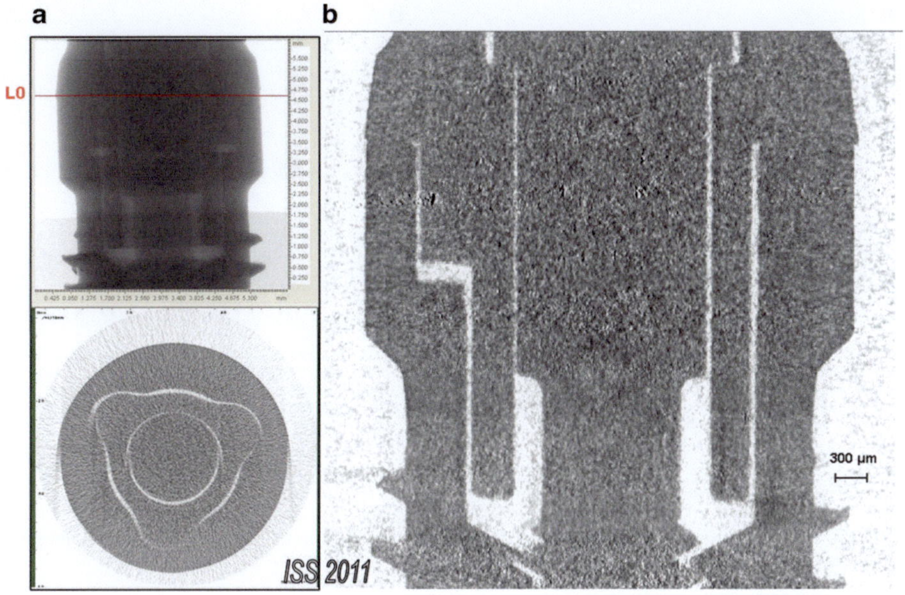

Fig. 15 Microtomographic horizontal (**a**) and vertical (**b**) section of Sample 2 with its Rx image (*left corner*) and indication of the L0 level

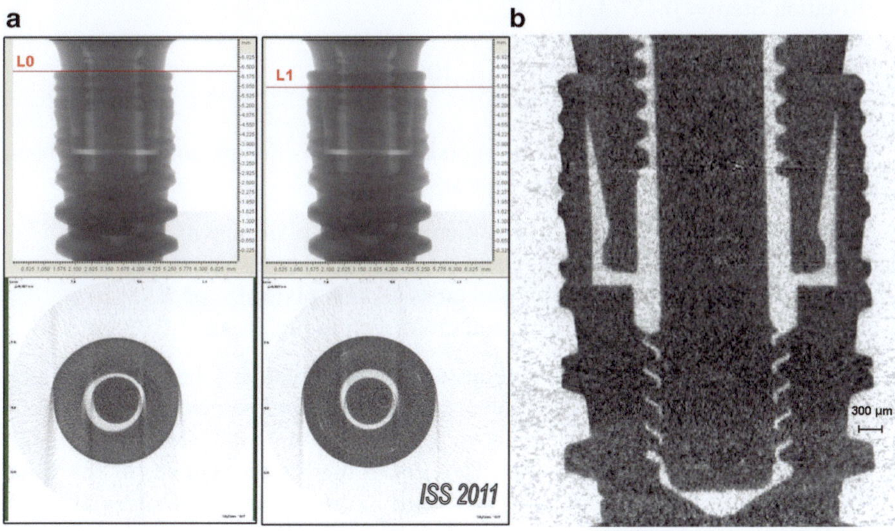

Fig. 16 Microtomographic horizontal (**a**) and vertical (**b**) section of Sample 3 with its Rx images (*left corner*) and indication of the L0, L1 levels

the accuracy of the tightness seal of the chosen connections (Di Carlo et al. 2008; Bedini et al. 2007).

Every sample has been glued in resin in vertical position inside a cylinder shaped mould in order to be acquired without moving artefacts in the microtomographic instrument.

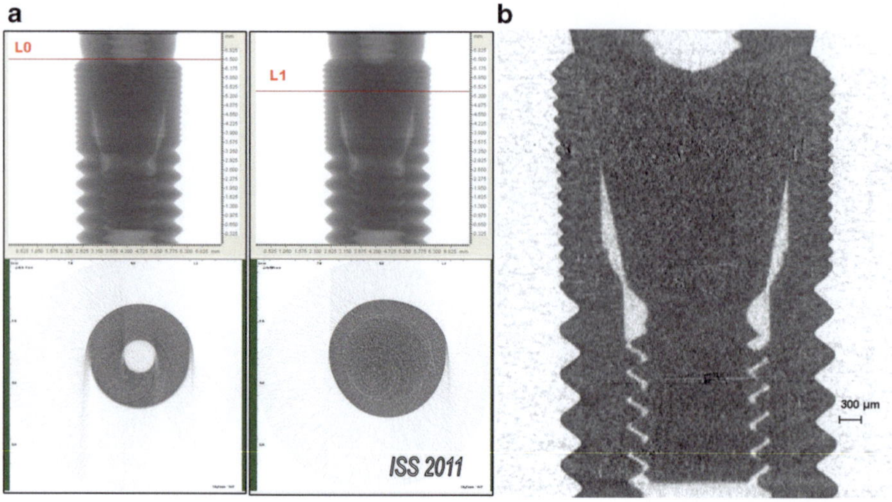

Fig. 17 Microtomographic horizontal (**a**) and vertical (**b**) section of Sample 4 with its Rx images (*left corner*) and indication of the L0, L1 levels

By software, a sample has been placed moving in vertical or horizontal way to focus the selected area and then acquire according to the following parameters:

- Rotation Step = 0.45° ± 0.01°C,
- Total rotation angle = 180° ± 1°C,
- Power source 100KV / 98 microA ± 10%
- Filter thickness 1 ± 0.01 mm (Al)

The magnification and cross-section pixel size acquisition parameters have been chosen according to the following values:

- Sample 1: magnification at 26 X and cross section pixie size of 11.27 um
- Sample 2: magnification at 50 X and cross section pixie size of 5.86 um
- Sample 3: magnification at 40X and cross section pixie size of 7.32 um
- Sample 4: magnification at 40 X and cross section pixie size of 7.32 um

All the images obtained from microtomographic scanning have been treated with a dedicated software, able to reconstruct a true three-dimensional model of selected implants, processing them in all external-internal features, through acquired sections, with no need to destroy, cut or modify the samples in any way.

Once a sample has been reconstructed with a certain number of slices of about 600–900, is possible to individuate and analyze the fixture-abutments' contact zones to observe the seal of the different connections sites.

About flat/horizontal connections, the axial section at level of fixture platform has been identified (L0, Figs. 1 and 2). By means of CTan software it has been possible to measure the surface of circular section corresponding to the contact between fixture platform and the base of the abutment in sample 1 and that of internal tri-channel connection observed in sample 2.

Table 3 Results of microtomographic analysis of fixture-abutment connections of four processed dental implant systems

Fixture-abutment contact surface				
Sample	Height of contact (mm)	Major radius (mm)	Minor radius (mm)	Contact surface (mm^2)
1	Not available	2.43 ± 0.02	1.57 ± 0.03	10.78 ± 0.04
2	Not available	2.44 ± 0.02	Not available	11.05 ± 0.03
3	0.53 ± 0.03	1.73 ± 0.03	1.62 ± 0.04	6.31 ± 0.03
4	1.11 ± 0.04	1.60 ± 0.01	1.39 ± 0.03	11.63 ± 0.04

In conical connections, through the sequential analysis of all reconstructed axial sections, it has been decided that L0 identifies the initial contact between implant and abutment, while L1 identifies the section in which can be observed a micro-gap between the two near surfaces, at the end of connection's seal (Figs. 3 and 4). By means of CTan software it has been also possible to measure the lateral surface of truncated cone between L0 and L1 that indicates the contact surface between the two components.

2.3 *Results*

Table 3 shows mean values of contact areas for each fixture-abutment connection obtained by microtomographic analysis with also the minor and major radius together with the height of contact for sample 3 and 4connections.

It's possible to observe how sample 3 contact area is very smaller than the others, even if in the sample 1 and 2 the values are similar, while in sample 3 and 4 are very different.

2.4 *Discussion*

The presence of a misfit between fixture and abutment is well known as may cause an increase of the mechanical stress on the dental implant system connection structures and also on the surrounding bone tissue. This condition can be the cause of a pre-load loss or fracture of tighten screws but also of some biological consequences (Albrektsson 1988; Oh et al. 2002; Ricomini Filho et al. 2010; Broggini et al. 2006).

A favorable clinical outcome of a dental implant is due to an appropriate osseointegration process and then to a long term integrity of the surrounding bone support. The presence of a micro-gap in the fixture-abutment interface, caused by a misfit of connection structures, allows the penetration and colonization of bacteria in the most internal side of the implant (Oh et al. 2002; Ricomini Filho et al. 2010;

Gratton et al. 2001; Broggini et al. 2006; Hecker and Eckert 2003; Cibirka et al. 2001; Berglundh et al. 2007).

Many authors wrote about the importance of fixture-abutment interface conditions on the level and the loss of surrounding bone tissue (Ricomini Filho et al. 2010; Broggini et al. 2006). In 2006, some researchers demonstrated that bone or peri-implant gingival tissue near the micro-gap can be subjected to inevitable inflammatory process (Broggini et al. 2006). As previously said, the micro gap allows the penetration of bacteria inside the fixture-abutment structure and allows also the endo-toxins produced in this internal structure to move from internal to external level through the surrounding tissues. In this case, a physio-pathological process can be triggered while, in the worst case, there could be bone resorption and loss of the whole implant system (Rack et al. 2010; Yi et al. 2001; Hermann et al. 2001; Jansen et al. 1997; Steinebrunner et al. 2005).

Due to that, different innovative implant system focused their attention on connections' design to avoid peri-implant tissue inflammation. Conic connection interfaces without tighten screws, as conic Morse taper and locking taper, showed a perfect seal connection between fixture and abutment in order to avoid bacteria penetration and any kind of mechanical problem (Ricomini Filho et al. 2010; Mangano et al. 2009; Dibart et al. 2005; Norton 1999). Furthermore, conic connections shows an interface right in the middle of the fixture internal platform, differently from the hexagon external connections, where the interface is close to the alveolar bone (Ricomini Filho et al. 2010; Mangano et al. 2009; Weng et al. 2008).

The necessity to obtain a good seal against bacteria infiltration produced a lot of studies on micro organisms penetration through micro-gaps of implant-abutment interface. Most of these studies showed the problem of an in vitro and in static condition seal, avoiding to consider important factors as temperature variations or chewing stresses, typical of human oral cavity (Ricomini Filho et al. 2010; Mangano et al. 2009; Persson et al. 1996; Gross et al. 1999; Guindy et al. 1998; Covani et al. 2006; Quirynen et al. 1994).

In the last decade, a lot of industries tried to realize a better connection between the different components of an implant system. At this moment, conic connection represents the best bacteria sealing due to their good fixture-abutment fitting, anyway there are several studies demonstrating that the ideal implant connection hasn't been reached yet. This means that the risk of a bacteria infiltration can't be totally avoided still today (Ricomini Filho et al. 2010; Mangano et al. 2009; Gross et al. 1999; Covani et al. 2006; Quirynen et al. 1994; Piattelli et al. 2001; Rimondini et al. 2001).

After the analysis of data showed in Table 2 it's possible to do two observations.

First of all, the connection's geometry influences the mechanical properties of an implant system. Two flat surfaces in contact show less possibilities to distribute occlusal loading, especially eccentric ones, in an homogeneous and multi directional way respect to another connection as the conometric one that has, on the other hand, contact surface characterized also by a vertical component inside implant body. Therefore, from a biomechanical point of view, a conic connection it's more geometrically suitable than a flat one, as well documented in literature

(Gratton et al. 2001; Mangano et al. 2009, 2010; Dibart et al. 2005; Binon 1996a, b, 2000; Coelho et al. 2008; Tsuge et al. 2008; Cibirka et al. 2001; Norton 1999, 2000; Piattelli et al. 2001; Watson 1998; Harder et al. 2009; Weng et al. 2010). In addition to that, it's possible to observe that, instead of problems connected to chewing load on surrounding bone tissue, at bone crest level, there are other serious problems as mechanical stress on the prosthetic component of the implant support that results more stressed flat connections. This kind of connection, in fact, is often subjected to mechanical stress as unscrewing or fracture of tighten screw, abutment or fixture fracture in the worst cases (Dibart et al. 2005; Patterson and Johns 1992; Di Carlo et al. 2008; Bedini et al. 2007; Gratton et al. 2001; Tsuge and Hagiwara 2009).

At the end, analyzing the numerical value of the contact surfaces, sample n 3, in spite of a conic connection, shows a fixture-abutment contact surface smaller than the one of the other three samples, each of them showing, on the other hand, very similar values after comparison.

2.5 Conclusions

The fitting of mechanical connections of implant components has a huge importance in peri-implant pathology prevention. The created fixture-abutment seal and the extension of the contact surface of the same one represent a very important starting point in preventing two of the most important kind of implant failures: inflammation of peri-implant tissues, due to bacteria penetration trough the gap between fixture and abutment structure, and the mechanical failure of one of the components of the implant system, due to a misfit between the fixture and the abutment, that amplifies the transfer of chewing loads on the same structure (Albrektsson 1988; Oh et al. 2002; Quirynen et al. 2002; Ricomini Filho et al. 2010; Gratton et al. 2001; Mangano et al. 2009; Dibart et al. 2005; Weng et al. 2008; Persson et al. 1996; Binon 1996a, b, 2000; Brunski 1995).

Nowadays, commercial development allows clinic to choose between several implant systems. In spite of the high number on the market, literature shows today that tube in tube conic shape of fixture-abutment contact has a better seal against bacteria and a better mechanical stability (Ricomini Filho et al. 2010; Gratton et al. 2001; Mangano et al. 2009, 2010; Dibart et al. 2005; Binon 1996b, 2000; Brunski 1995; Coelho et al. 2008; Tsuge et al. 2008; Cibirka et al. 2001; Norton 1999, 2000; Gross et al. 1999; Covani et al. 2006; Quirynen et al. 1994; Piattelli et al. 2001; Rimondini et al. 2001; Watson 1998; Harder et al. 2009; Weng et al. 2010).

From the point of view of designing and developing dental implant systems, which become increasingly effective and reliable, the technology applied to the production of implant-prosthetic components is of great importance. For this aim, the use of X-ray microtomography works as one of the best tool in this kind of applications, comparing it to traditional investigation methods, because allows to perform three dimensional images and evaluations in a non invasive and non destructive way.

Acknowledgments The authors wish to thank professors L. Pacifici and L. Baggi for clinical support and professor G. Soda for generously supplying the histological experimental tests.

References

Albrektsson T (1988) A multicenter report on osseointegrated oral implants. J Prosthet Dent 60:75–84

Bedini R, Ioppolo P, Pecci R, Rizzo F, Di Carlo F, Quaranta M (2007) Studio in vitro sulla connessione di sistemi implantari dentali. Rapporti ISTISAN 07/7

Berglundh T, GotfredsenK ZNU, Lang NP, Lindhe J (2007) Spontaneous progression of ligature induced peri-implantitis at implants with different surface roughness: an experimental study in dogs. Clin Oral Implants Res 18:655–661

Bianchi F, Perrotti G, Francetti L, Testori T (2002) L'estetica in implantologia. Un caso clinico di agenesia dentale. Ital Oral Surg 1(1):41–46

Bickford J Jr (1981) An introduction to the design and behavior of bolted joints. Marcel Decker, New York

Binon PP (1996a) The effect of eliminating implant/abutment rotational misfit on screw joint. Int J Prosthodont 9(6):511–519

Binon PP (1996b) The spline implant: design, engineering, and evaluation. Int J Prosthodont 9 (5):419–433

Binon PP (2000) Impianti e componenti all'alba del nuovo millennio. Quintess Int 9/10:317–330

Bragger U (1998) Use of radiographs in evaluating success, stability and failure in implant dentistry. Periodontol 2000 17:77–88

Broggini N, McManus LM, Hermann JS, Medina R, Schenk RK, Buser D (2006) Peri-implant inflammation defined by the implant-abutment interface. J Dent Res 85:473–478

Brunski JB (1995) Biomechanics of dental implants. In: Block MS, Kent JN (eds) Endosseus implants for maxillofacial reconstruction. Sounders, Philadelphia

Burchardt H (1983) The biology of bone graft repair. Clin Orthop Relat Res 174:28–42

Cancedda R, Cedola A, Giuliani A, Komlev V, Lagomarsino S, Mastrogiacomo M, Peyrin F, Rustichelli F (2007) Bulk and interface investigations of scaffolds and tissue-engineered bones by X-ray microtomography and X-ray microdiffraction. Biomaterials 28(15):2505–2524 (Review)

Chiapasco M, Romeo E (2002) La riabilitazione implantoprotesica nei casi complessi. UTET, Milano

Chu TM, Warden SJ, Turner CH, Stewart RL (2007) Segmental bone regeneration using a load-bearing biodegradable carrier of bone morphogenetic protein-2. Biomaterials 28(3):459–467

Cibirka RM, Nelson SK, Lang BR, Rueggeberg FA (2001) Examination of the implant-abutment interface after fatigue testing. J Prosthet Dent 85:268–275

Coelho AL, Suzuki M, Dibart S, Da Silva N, Coelho PG (2007) Cross-sectional analysis of the implant-abutment interface. J Oral Rehabil 34:508–516

Coelho PG, Sudack P, Suzuki M, Kurtz KS, Romanos GE, Silva NRFA (2008) In vitro evaluation of the implant abutment connection sealing capability of different implant systems. J Oral Rehabil 35:917–924

Covani U, Marconcini S, Crespi R, Barone A (2006) Bacterial plaque colonization around dental implant surfaces. Implant Dent 15:298–304

Cowan CM, Aghaloo T, Chou YF, Walder B, Zhang X, Soo C, Ting K, Wu B (2007) MicroCT evaluation of three-dimensional mineralization in response to BMP-2 doses in vitro and in critical sized rat calvarial defects. Tissue Eng 3(3):501–512

Di Carlo F, Marincola M, Quaranta A, Bedini R, Pecci R (2008) Analisi MicroTac di impianti a connessione conometrica. Dent Cadmos 76(3):1–6

Dibart S, Warbington M, Su MF, Skobe Z (2005) In vitro evaluation of the implant-abutment bacterial seal: the locking taper system. Int J Oral Maxillofac Implants 20:732–737
Feldkamp LA, Davis LC, Kress JW (1984) Practical cone-beam algorithm. J Opt Soc Am A 1:612–619
Feldkamp LA, Goldstein SA, Parfitt AM, Jesion G, Kleerekoper M (1989) The direct examination of three-dimensional bone architecture in vitro by computed tomography. J Bone Miner Res 4 (1):3–11
Freed LE, Guilak F, Guo XE, Gray ML, Tranquillo R, Holmes JW, Radisic M, Sefton MV, Kaplan D, Vunjak-Novakovic G (2006) Advanced tools for tissue engineering: scaffolds, bioreactors and signaling. Tissue Eng 12:3285–3305
Gielkens PF, Schortinghuis J, de Jong JR, Raghoebar GM, Stegenga B, Bos RR (2008) Vivosorb, Bio-Gide, and Gore-Tex as barrier membranes in rat mandibular defects: an evaluation by microradiography and micro-CT. Clin Oral Implants Res 19(5):516–521
Gratton DG, Aquilino SA, Stanford CM (2001) Micromotion and dynamic fatigue properties of the dental implant-abutment interface. J Prosthet Dent 85:47–52
Gross M, Abramovich I, Weiss EI (1999) Microleakage at the abutment-implant interface of osseointegrated implants: a comparative study. Int J Oral Maxillofac Implants 14:94–100
Guindy JS, BesimoCE BR, Schiel H, Meyer J (1998) Bacterial leakage into and from prefabricated screw-retained implant-borne crowns in vitro. J Oral Rehabil 25:403–408
Guldberg RE, Duvall CL, Peister A, Oest ME, Lin AS, Palmer AW, Levenston ME (2008) 3D imaging of tissue integration with porous biomaterials. Biomaterials 29(28):3757–3761 (Epub 2008 Jul 16)
Haïat G, Padilla F, Peyrin F, Laugier P (2007) Variation of ultrasonic parameters with microstructure and material properties of trabecular bone: a 3D model simulation. J Bone Miner Res 22 (5):665–674
Harder S, Dimaczek B, Acil Y, Terheyden H, Freitag-Wolf S, Kern M (2009) Molecular leakage at implant-abutment connection – in vitro investigation of tightness of internal conical implant-abutment connections against endotoxin penetration. Clin Oral Invest 14(4):427–432
Hecker DM, Eckert SE (2003) Cyclic loading of implant-supported prostheses: changes in component fit over time. J Prosthet Dent 89:346–351
Hecker DM, Eckert SE, Choi YG (2006) Cyclic loading of implant-supported prostheses: comparison of gaps at the prosthetic-abutment interface when cycled abutments are replaced with as-manufactured abutments. J Prosthet Dent 95:26–32
Hermann JS, Schoolfield JD, Schenk RK, Buser D, Cochran DL (2001) Influence of the size of the microgap on crestal bone changes around titanium implants. A histometric evaluation of unloaded non-submerged implants in the canine mandible. J Periodontol 72:1372–1383
Hofmann S, Hagenmuller H, Koch AM, Muller R, Vunjak-Novakovic G, Kaplan DL, Merkle HP, Meinel L (2007) Control of in vitro tissue-engineered bone-like structures using human mesenchymal stem cells and porous silk scaffolds. Biomaterials 28(6):1152–1162
Jansen VK, Conrads G, Richter EJ (1997) Microbial leakage and marginal fit of the implant-abutment interface. Int J Oral Maxillofac Implants 12:527–540
Jones JR, PoologasundarampillaiG ARC, Bernard D, Lee PD (2007) Non-destructive quantitative 3D analysis for the optimisation of tissue scaffolds. Biomaterials 28(7):1404–1413
Jorneus L, Jemt T, Carlsson L (1992) Loads and designs of screw joints for single crowns supported by osseointegrated implants. Int J Oral Maxillofac Implants 7:353–359
Kachelrieb M (2008) Micro-CT. Handb Exp Pharmacol 185(1):23–52 (Review)
Kamburoglu K, Barenboim SF, Aritürk T, Kaffe I (2008) Quantitative measurements obtained by micro-computed tomography and confocal laser scanning microscopy. Dentomaxillofac Radiol 37(7):385–391
Kerckhofs G, Schrooten J, Van Cleynenbreugel T, Lomov SV, Wevers M (2008) Validation of x-ray microfocus computed tomography as an imaging tool for porous structures. Rev Sci Instrum 79(1):013711

Komlev VS, Peyrin F, Mastrogiacomo M, Cedola A, Papadimitropoulos A, Rustichelli F, Cancedda R (2006) Kinetics of in vivo bone deposition by bone marrow stromal cells into porous calcium phosphate scaffolds: an X-ray computed microtomography study. Tissue Eng 12(12):3449–3458

Lin-Gibson S, Cooper JA, Landis FA, Cicerone MT (2007) Systematic investigation of porogen size and content on scaffold morphometric parameters and properties. Biomacromolecules 8(5):1511–1518 (Epub 2007 Mar 24)

Mangano C, Mangano F, Piattelli A, Iezzi G, Mangano A, Lacolla L (2009) Prospective clinical evaluation of 1920 Morse taper connection implants: results after 4 years of functional loading. Clin Oral Implants Res 20:254–261

Mangano C, Mangano F, Piattelli A, Iezzi G, Mangano A, La Colla L (2010) Prospective clinical evaluation of 307 single-tooth morse taper-connection implants: a multicenter study. Int J OralMaxillofac Implants 25:394–400

Maréchal M, Luyten F, Nijs J, Postnov A, Schepers E, van Steenberghe D (2005) Histomorphometry and micro-computed tomography of bone augmentation under a titanium membrane. Clin Oral Implants Res 16(6):708–714

Mastrogiacomo M, Muraglia A, Komlev V, Peyrin F, Rustichelli F, Crovace A, Cancedda R (2005) Tissue engineering of bone: search for a better scaffold. OrthodCraniofac Res 8 (4):277–284

McGlumpy EA, Mendel DA, Holloway JA (1998) Implant screw mechanics. Dent Clin North Am 42:71–89

Norton MR (1999) Assessment of cold welding properties of the internal conical interface of two commercially available implant systems. J Prosthet Dent 81:159–166

Norton MR (2000) An in vitro evaluation of the strength of a 1-piece and 2-piece conical abutment joint in implant design. Clin Oral Implants Res 11:458–464

Oh TJ, Yoon J, Misch CE, Wang HL (2002) The causes of early implant bone loss: myth or science? J Periodontol 73:322–333

Papadimitropoulos A, Mastrogiacomo M, Peyrin F, Molinari E, Komlev VS, Rustichelli F, Cancedda R (2007) Kinetics of in vivo bone deposition by bone marrow stromal cells within a resorbable porous calcium phosphate scaffold: an X-ray computed microtomography study. Biotechnol Bioeng 98(1):271–281

Parkinson IH, Badiei A, Fazzalari NL (2008) Variation in segmentation of bone from micro-CT imaging: implications for quantitative morphometric analysis. Australas Phys Eng Sci Med 31 (2):160–164

Patterson EA, Johns RB (1992) Theoretical analysis of the fatigue life of fixture screws in osseointegrated dental implants. Int J Oral Maxillofac Implants 7:26–33

Perilli E, Baruffaldi F, Visentin M, Bordini B, Traina F, Cappello A, Viceconti M (2007) MicroCT examination of human bone specimens: effects of polymethylmethacrylate embedding on structural parameters. J Microsc 225(Pt 2):192–200

Persson LG, Leckholm U, Leonhardt A, Dahlen G, Lindhe J (1996) Bacterial colonization on internal surfaces of Branemark system implant components. Clin Oral Implants Res 7:90–95

Piattelli A, Scarano A, Paolantonio M, Assenza B, Leghissa GC, Di Bonaventura G (2001) Fluids and microbial penetration in the internal part of cemented-retained versus screw-retained implant-abutment connections. J Periodontol 72:1146–1150

Quirynen M, BollenCM EH, van Steenberghe D (1994) Microbial penetration along the implant components of the Branemark system. An in vitro study. Clin Oral Implants Res 5:239–244

Quirynen M, De Soete M, van Steenberghe D (2002) Infectious risks for oral implants: a review of the literature. Clin Oral Implants Res 13:1–19

Rack A, Rack T, Stiller M, Riesemeier H, Zabler S, Nelson K (2010) In vitro synchrotron-based radiography of micro-gap formation at the implant-abutment interface of two-piece dental implants. J Synchrotron Radiat 17:289–294

Ricomini Filho AP, FernandesFSdeF SFG, da Silva WJ, Del BelCury AA (2010) Preload loss and bacterial penetration on different implant-abutment connection systems. Braz Dent J 21 (2):123–129

Rimondini L, Marin C, Brunella F, Fini M (2001) Internal contamination of a 2-component implant system after occlusal loading and provisionally luted reconstruction with or without a washer device. J Periodontol 72:1652–1657
Stauber M, Müller R (2008) Micro-computed tomography: a method for the non-destructive evaluation of the three-dimensional structure of biological specimens. Methods Mol Biol 455:273–292
Steinebrunner L, Wolfart S, Bossmann K, Kern M (2005) In vitro evaluation of bacterial leakage along the implant-abutment interface of different implant systems. Int J Oral Maxillofac Implants 20:875–881
Tsuge T, Hagiwara Y (2009) Influence of lateral-oblique cyclic loading on abutment screw loosening of internal and external hexagon implants. Dent Mater J 28(4):373–381
Tsuge T, Hagiwara Y, Matsamura H (2008) Marginal fit and microgaps of implant-abutment interface with internal anti-rotation configuration. Dent Mater J 27(1):29–34
Urist MR (1965) Bone: formation by autoinduction. Science 150:893–899
Urist MR, Sato K, Brownell AG, Malinin TI, Lietze A, Huo YK, Prolo DJ, Oklund S, Finerman GA, DeLange RJ (1983) Human bone morphogenetic protein (hBMP). Proc Soc Exp Biol Med 173(2):194–199
van Lenthe HG, Hagenmuller H, Bohner M, Hollister SJ, Meinel L, Muller R (2007) Nondestructive micro-computed tomography for biological imaging and quantification of scaffold-bone interaction in vivo. Biomaterials 28:2479–2490
Waarsing JH, Day JS, Weinans H (2004) An improved segmentation method for in vivo microCT imaging. J Bone Miner Res 19(10):1640–1650 (Epub 2004 Jul 12)
Watson PA (1998) Sviluppo e produzione delle componenti protesiche:c'è bisogno di cambiamenti? Int J Prosthodont 11:513–516
Weng D, Nagata MJ, Bell M, Bosco AF, de Melo LG, Richter EJ (2008) Influence of microgap location and configuration on the periimplant bone morphology in submerged implants. An experimental study in dogs. Clin Oral Implants Res 19:1141–1147
Weng D, Nagata MJH, Bell M, de Melo LGN, Bosco AF (2010) Influence of mcrogap location and configuration on peri-implant bone morphology in nonsubmerged implants: an experimental study in dogs. Int J Oral Maxillofac Implants 25(3):540–547
Willems NM, Mulder L, Langenbach GE, Grunheid T, Zentner A, van Eijden TM (2007) Age-related changes in microarchitecture and mineralization of cancellous bone in the porcine mandibular condyle. J Struct Biol 158(3):421–427 (Epub 2007 Jan 8)
Yi JM, Lee JK, Um HS, Chang BS, Lee MK (2001) Marginal bony changes in relation to different vertical positions of dental implants. J Periodontal Implant Sci 40:244–248
Yip G, Schneider P, Roberts EW (2004) Micro-computed tomography: high resolution imaging of bone and implants in three dimensions. Semin Orthod 10:174–187
Zipprich H, Weigl P, Lange B, Lauer HC (2007) Erfassung, Ursachen und Folgen von Mikrobewegungen am Implantat-Abutment-Interface. Implantologie 15:31–46

Biomedical Imaging and Computational Modeling in Cardiovascular Disease: Patient-Specific Applications Using Numerical Models

Vanessa Díaz-Zuccarini and Silvia Schievano

Abstract Patient-specific simulations will become in the not so distant future a "de facto" standard in surgical planning and diagnostic. Nowadays, in-silico simulations have the power to explore in detail a system close to its "in-vivo" conditions, and to report values difficult or impossible to characterize due to physical or ethical constraints. The objective of this chapter is to present two applications of patient-specific simulations for heart valves: In Application 1, a coupled 3D-1D of a mechanical heart valve is shown. In application 2, a structural patient-specific study for percutaneous valve implantations is presented.

Keywords Cardiovascular disease • Computational analysis • Imaging

1 Application 1: Computational Models and Appropriate Patient-Specific Boundary Conditions for CFD Models

If it were possible to simulate the entire human body, one could envisage a completely closed fluid model, including the heart, arteries, capillaries and veins. However, to model this as a single three-dimensional model is not only impractical, is clearly impossible. For this reason, the patient scan data required to obtain the geometry is often confined to a particular region of the body. This means that, at some point, the fluid model must be truncated to give inlets and outlets where

V. Díaz-Zuccarini (✉)
Department of Mechanical Engineering, University College London, Malet Place Engineering Building, Torrington Place, WC1E 7JE London, UK
e-mail: v.diaz@ucl.ac.uk

S. Schievano
Institute of Cardiovascular Science and Great Ormond Street Hospital for Children, University College London, London, UK
e-mail: s.schievano@ich.ucl.ac.uk

D. Iacoviello and U. Andreaus (eds.), *Biomedical Imaging and Computational Modeling in Biomechanics*, Lecture Notes in Computational Vision and Biomechanics 4,
DOI 10.1007/978-94-007-4270-3_9,

boundary conditions must be specified. In the simplest case, flow rates or pressures are imposed directly at the boundaries. Alternatively, a representation of the remainder of the circulatory system can be coupled to the boundaries. At its least complex, the model could be a simple resistance or a Windkessel-type model. The problem of coupling together models of differing complexity has been considered by (Formaggia et al. 2001). They coupled a 3D fluid model with a solid model wall to a non-linear 1D model at the outlet. This was found to be an effective way of reducing spurious wave reflections at the outlet boundary of the 3D model. The coupling of a 3D model to a 0D compartment model, using state variables to describe the compartment model has been analysed by (Quarteroni and Veneziani 2003). This approach in the clinical application of paediatric cardiac surgery is used by (Migliavacca et al. 2005). Also, a scheme for coupling a 3D model to a digital implementation of the Westerhof model (both, in frequency and time domain) (Westerhof et al. 1969) has been implemented. These "realistic" representations of the arterial tree for patient-specific simulations (Jones et al. 2004) are of paramount importance in cardiovascular engineering and clinically-meaningful simulations. The appropriate description of boundary conditions for the arterial load, using patient specific data (length of vessels, diameter, etc.) is possible nowadays. However, this is just the start. The cardiovascular system is a coupled heart-arterial network system, thus both the pumping and load systems must be described correctly. Dynamic boundary conditions, described in terms of pressure and flow, are certainly better than imposed input and output conditions from a haemodynamic point of view. This section, illustrates how these ideas have been developed further.

1.1 Materials and Methods

A coupled system based on previously published material (Díaz-Zuccarini et al. 2008) of a mechanical heart valve (MHV) and biologically-based boundary conditions of the pumping-system is briefly described (Rafiroiu et al. 2008), taking into account fundamental biochemical details of cardiac contraction. One possible application of this methodology is the exploration of the effect of pharmacological intervention on the behaviour of medical devices and implants. This is a key issue since patients undergoing this type of surgery normally receive medication for life; information could be provided on possible outcomes and alternative treatments. Although the results presented in this chapter are reported on an "idealised" valve, laser scanning and imaging techniques can be used in order to simulate commercially available valves, as shown on Fig. 1 (Rafiroiu et al. 2008).

This section describes the development of more complex boundary conditions to represent the behaviour of the left ventricle (LV).

Figure 2 shows the sub-levels of organisation of the ventricle that are taken into account to give a "more physiologically realistic" model of the LV. Following the procedure described in *ex vivo* isolated LV preparations, a constant pressure source is connected to the ventricle via the mitral valve (input model). The blood fills the LV (output model) and ejects a volume of blood into the arterial network via the

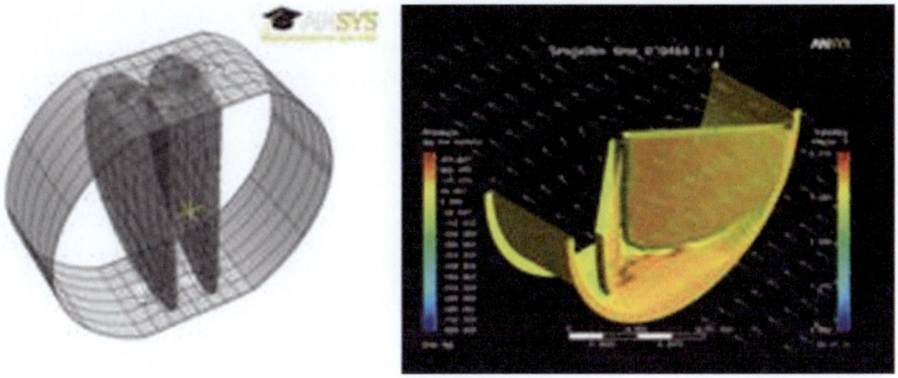

Fig. 1 Scanned "St Jude" valve and preliminary simulation results. There is a symmetry axis, i.e., the valve is "cut in half" to reduce the simulation time

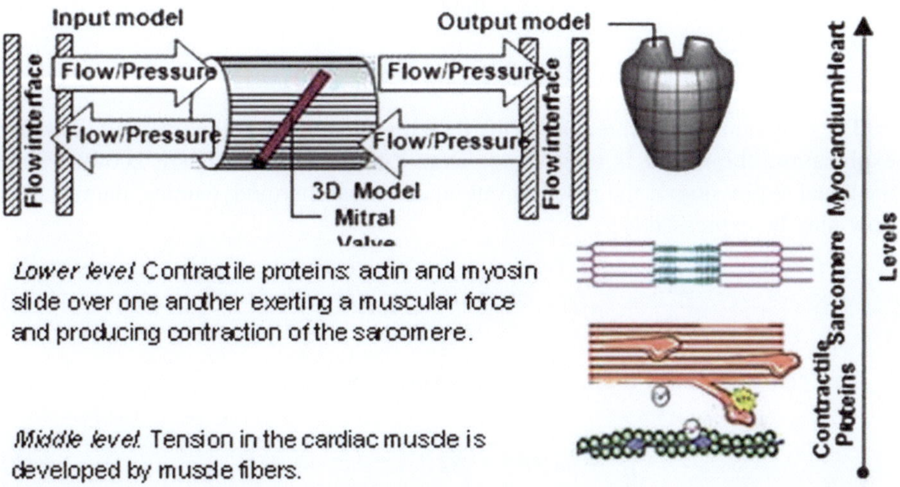

Fig. 2 Representation of the LV model and its different physical scales. Cardiac contraction starts at protein level in the sarcomere. Protein contraction produces tension in the cardiac wall that is then translated into pressure

aortic valve. Contraction in the cardiac muscle is described starting from the microscopic level of contractile proteins (actin and myosin), up to the tissue (muscle level) and to the LV level to finally reach the haemodynamic part of the LV and its arterial load.

1.2 A Multi-Scale Model of the Left Ventricle

The model published in (Díaz-Zuccarini et al. 2008, 2009) provides the basis for this coupling. The fundamental mechanisms of cardiac contraction start at the protein scale. The contractile proteins (actin and myosin) acting at the level of the sarcomere produce cardiac contraction by sliding over each other, attaching and detaching by means of cross-bridges, to produce shortening of the sarcomere and hence contraction of the ventricular wall (Fig. 2). To represent the chemical reaction from the detached state (X_D) to the attached one (X_A) and vice-versa, the simplest model available was used:

$$X_D \underset{k_d}{\overset{k_a}{\rightleftarrows}} X_A \tag{1}$$

The periodically varying kinetic constants of attachment and detachment are k_a(t) and k_d(t).

A modified Nernst formula provides a first approximation for the calculation of the chemical potential based on the "attached" X_A previously described:

$$\mu_a = (A_a + B_a(T)\ln(X_A))\beta(l_m) \tag{2}$$

$$\mu_d = A_d + B_d(T)\ln(X_D) \tag{3}$$

The only non-classical feature is the β factor of equation (2). For a more detailed explanation, the reader is referred to (Díaz-Zuccarini and LeFèvre 2007). Using the usual expression of the reaction rate in chemical thermodynamics, the reaction flow may be expressed as:

$$\frac{dX_A}{dt} = k_A(t) \cdot e^{(\mu_d - A_d)/B_d} - k_D(t) \cdot e^{(\mu_a - \beta A_a)/Ba} \tag{4}$$

This can be reduced to the following expression:

$$\frac{dX_A}{dt} = k_A(t)(1 - X_A(t)) - k_D(t)X_D(t)^{\beta} \tag{5}$$

In order to describe the *chemo-mechanical* transduction in the cardiac muscle, both chemical and mechanical aspects of the model must be interlinked. A single muscle equivalent to all the muscles of the heart is considered. Its length is l_m and X_A is its number of attached cross bridges. Neglecting the energy in free cross bridges, the energy of the muscle is given by:

$$E_m = E_m(l_m, X_A) \tag{6}$$

Mechanical power is the derivative of energy with respect to time. The mechanical power is then obtained by derivation:

$$P_m = \frac{\partial E_m}{\partial l_m}\frac{dl_m}{dt} + \frac{\partial E_m}{\partial X_a}\frac{dX_A}{dt} = f_m v_m + \mu_a \dot{X}_A \tag{7}$$

Where f_m is the force of the ventricular muscle and v_m is the velocity.

As shown in Eq. 7, the mechanical power is obtained by derivation of the energy equation of the muscle with respect to the attached cross-bridges (E_X) and the length of muscle fibres (l_m).

$$f_m = E_X(X_A)E'_a(l_m) + (1 - E_X(X_A))E'_r(l_m) \tag{8}$$

E'_a and E'_r are respectively, the derivative of the energy of the muscle in active (a) and resting (r) state with respect to l_m and they are expressed as functions of the muscular force, f_m. Full details of the formulation and assumptions may be found in (Díaz-Zuccarini and LeFèvre 2007). In our model, the cardiac chamber transforms muscular force into pressure and volumetric flow into muscle velocity. We assume an empiric relationship φ_G between volume Q_{LV} and equivalent fibre length l_m

$$Q_{LV}(t) = \varphi_G(l_m(t)) \leftrightarrow l_m(t) = \varphi_G^{-1}(Q_{LV}(t)) \tag{9}$$

Assuming φ_G is known and by differentiation of (9), a relationship between the velocity of the muscle dl_m/dt and the ventricular volumetric flow dQ_{LV}/dt is obtained:

$$\frac{dQ_{lv}(t)}{dt} = \psi_G \frac{dl_m}{dt} \tag{10}$$

where $\psi_G = \psi_G(Q_{LV})$ is the derivative of φ_G with respect to l_m. If the transformation is power-conservative, mechanical power (depending on muscular force f_m) = hydraulic power (depending on pressure P).

$$P \cdot \frac{dQ_{lv}}{dt} = N \cdot f_m \cdot \frac{dl_m}{dt} \tag{11}$$

Where pressure P is assumed to be created from the forces of N identical fibres. Equation 12 is then obtained:

$$P = \frac{1}{\psi_G(l_m)} N f_m \tag{12}$$

In this case study, the ventricular pressure P defined (12) will be the input to the 3D model shown in the next section.

It is important to stress the fact that pressure P also depends on flow in the 3D model. The flow passing through the valve is calculated as the difference between the pressure produced by the ventricle and the pressure from the afterload (i.e., Windkessel model), divided by a resistance.

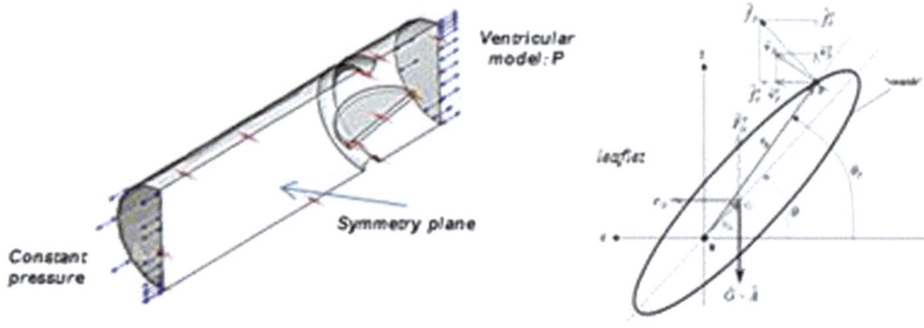

Fig. 3 *Left*: CAD model and boundary conditions: Snapshot of CFX-Pre implementation. At the inlet the ventricular pressure (P) is provided by the ventricular model. A constant pressure is applied at the outlet. *Right*: General representation of a rigid body demonstrating the principles of fluid-structure interaction and rotation of a leaflet. The variables are defined in the text

1.3 Computational Model of the Mitral Valve and Coupling with the LV Model

Mechanical heart valves are considered to be extremely durable in comparison to their bioprosthetic counterparts. The major load arises from transvalvular pressure generated at, and after, valve closure. In cases where structural failure does happen, it is usually as a result of occluder impact on the components. The estimation of closure forces on mechanical heart valves is, thus, of major interest as failure of the device can lead to catastrophic failure (Blackstone 2005; Prendergast and Maher 2001).

Cavitation is also an undesirable occurrence. Cavitation occurs when cavities or bubbles form in a liquid that is being pumped. These bubbles form at the low pressure or suction side of the pumping mechanism, causing the cavities or bubbles to collapse when they pass into the higher regions of pressure. Although the collapse of these cavities is a relatively low energy event, it is highly localized and can erode valve components causing erosion of material from the surface of the valve and surface pitting. Pits may act as stress-concentrators and, in some circumstances, may dramatically shorten a MHV's lifetime (Yoganathan et al. 2004).

The computational approach presented here is based on the use of a fluid-structure interaction model, built within ANSYS/CFX. This model describes the interaction of a blood-like fluid flowing in response to a pressure gradient imposed by ventricular contraction, and the structure of a single leaflet valve. For the purpose of the current case study, an idealized CAD model of a single-disc MHV, consisting of a flat leaflet (the occluder) which moves inside a mounting ring (the housing) was built to represent the geometry of the valve (Fig. 3). The thickness of the disc-shaped occluder is $h = 2$ mm and its diameter is $d = 21$ mm. In the fully closed position, the gap between the mounting ring and the occluder is $g = 0.5$ mm. The movement of the occluder is considered to be purely rotational, acting around an eccentric axis situated at 3.66 mm from its centroid (OX axis). The occluder travels 70° from the fully open to the fully closed position. To complete the CAD model, two cylindrical chambers were added to represent the atrium and ventricle. The atrial chamber, positioned in the positive OZ direction, is 66 mm in length whilst the ventricular chamber, positioned in the negative OZ direction, is 22 mm in length. The CAD model of the system, geometry details and boundary condition can be found in Fig. 3. Only half of the valve was considered in the 3D model. Blood is considered an incompressible Newtonian fluid with density $\rho = 1{,}100\ \text{kg/m}^3$ and dynamic viscosity $\mu = 0.004$ kg/ms.

The unsteady flow field inside the valve is described by the 3D equations of continuity and momentum (13) with the boundary conditions suggested by Fig. 3 (left)

$$\rho \frac{\partial \vec{u}}{\partial t} + (\vec{u} \cdot \nabla)\, \vec{u} \rho = -\nabla p + \mu \Delta \vec{u}; \quad \nabla \cdot \vec{u} = 0 \tag{13}$$

The principle used to model the fluid-structure interaction is outlined in Fig. 3 (right). The occluder rotates under the combined effects of hydrodynamic, buoyancy and gravitational forces acting on it. At every time step, the drag, $\vec{f_P^z}$, and lift, $\vec{f_P^y}$, forces exerted by the flowing fluid on an arbitrary point P of the leaflet's surface, are reduced to the centroid, G. The total contribution of drag, $\vec{F_G^z}$, and lift $\vec{F_G^y}$ are added to the difference between the gravitational and buoyancy forces. Conservation of the kinetic moment is imposed, resulting in the dynamic equation that describes the motion.

$$d\theta = \varpi_{old} dt + \frac{M}{J} dt^2 \tag{14}$$

In Eq. 14, $d\theta$ represents the leaflet's angular step, dt is the time step, ϖ_{old} is the angular velocity at the beginning of the current time step and M is the total momentum acting on the leaflet from the external forces:

$$M = r_G \left[F_G^z \sin\theta + (F_G^y - G + A)\cos\theta \right] \tag{15}$$

The moment of inertia of the occluder $J = 6.10^{-8}\ \text{kg.m}^2$.

To model the leaflet's rebound motion, the dynamic equation was modified (16) according to the following equation:

$$d\theta = \frac{\left[\varpi_{old} dt + \frac{M}{J} dt^2\right] + \frac{(\theta_{old} - \theta_{start}) C_{reb}}{J} dt^2}{1 - \frac{C_{reb}}{J} \delta t^2} \tag{16}$$

when, within the current time step, the new angular position drops below zero, a virtual torsional spring with constant $C_{reb} = 10$ Nm/degree is suddenly considered

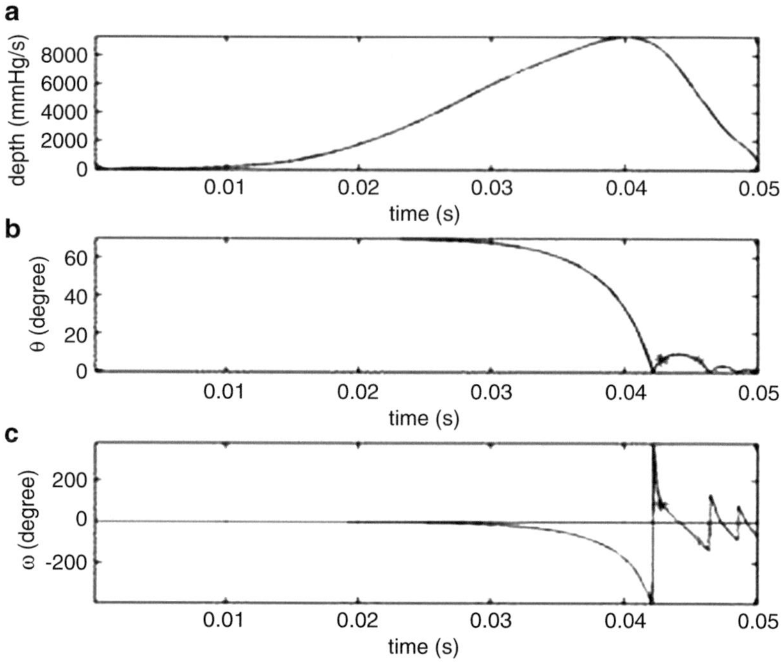

Fig. 4 (**a**) Rate of increase of ventricular pressure vs. time. (**b**) position history of the occluder vs. time. (**c**) Angular velocity of the valve vs. time

to be stretched between the fully opened ($\theta_{start} = 70°$) and the current position (θ_{old}); the elastic force developed by the spring is added to the gravitational force, thus reversing the leaflet's rotation:

1.4 Results and Discussion

Figure 4 shows the simulation results for a number of variables as a function of simulation time in seconds. These results have already been presented in (Díaz-Zuccarini et al. 2008). The rate of increase of ventricular pressure is shown in Fig. 4a, the change in ventricular pressure is determined by the definition of the boundary condition model described in Sect. 3.1. The pressure applied at the boundary depends on both the parameters used to define the boundary condition model and the flow calculated in the 3D model.

The angular position of the occluder is shown in Fig. 4b. A number of rebounds of the valve leaflet are observed after the initial closure event which is likely to be associated with the numerical implementation of valve closure, where the valve is restricted from rotating beyond an angular position of 0°. This leads to a large acceleration of the occluder at the instant when the valve reaches the 0° position, in the case of a *in vivo* valve this effect would be affected by the compliance of the valve mounting. This effect is even more noticeable if we consider the angular

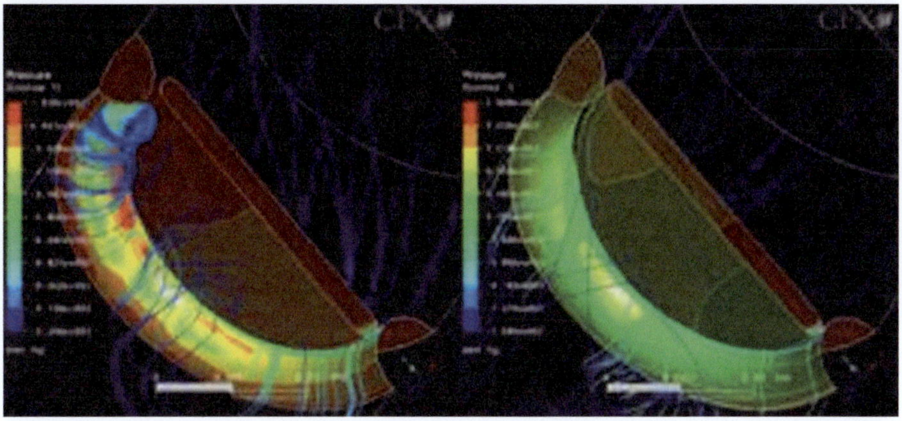

Fig. 5 *Left*: Pressure fields and vortex formation at the surface of the occluder and the housing, during rebound (at t = 0.0428 s). *Right*: Pressure fields at the surface of the occluder and the housing, during forward movement (at t = 0.0457 s)

velocity of the valve as shown in Fig. 4c. There is a significant discontinuity in the angular velocity at the instant of initial valve closure and at each successive time when the valve reaches an angular position of 0°.

It is interesting to note that from the implementation point of view, closure is continuous and smooth. The valve closure time predicted by the analysis is 34.4 ms, as measured from the moment when the valve starts to move until the end of the first rebound, which is within the range of experimentally reported values (Hose et al. 2006).

To investigate the cavitation potential of the valve, negative pressure values below the vaporization threshold and evidence of vortex formation on the surface of the leaflet were looked for. The pressure field in Fig. 5 clearly demonstrates the presence of negative gradients. Negative pressure values alternate on the surface of the leaflet, depending on the direction of movement (Fig. 5). During backward movement (rebound), negative pressure values and vortex flow predominate; whilst during forward movement, vortexes progressively change into swirl flow and negative pressure values decrease. This observation is consistent with experimental observations (Chandran et al. 1998) which, in the case of non-overlapping single-disc mechanical heart valves, (the Medtronic-Hall valve, for example) show that fluid vaporization is localized towards the upper (larger) gap.

2 Application 2: An Engineering Approach to Study Percutaneous Valve Implantation Using Structural Simulations

According to the American Heart Association, valvular heart disease is responsible for nearly 20,000 deaths each year in the United States, and is a contributing factor in about 42,000 deaths. Until recently, treatment has required surgical repair or

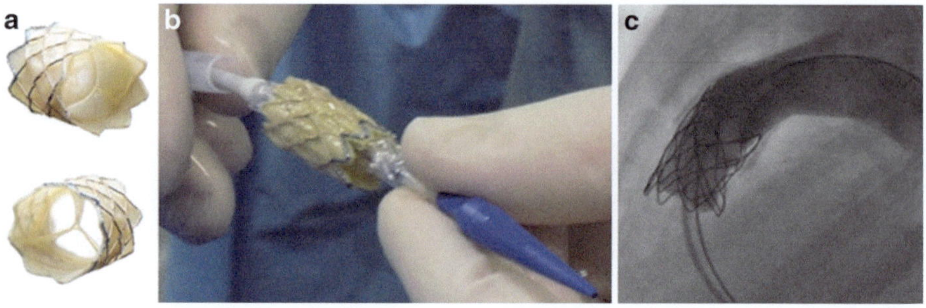

Fig. 6 The PPVI device (**a**) is mounted onto the catheter for delivery (**b**). Once in place, the device is inflated: the stent anchors the valve to the RVOT as shown from the post-PPVI angiogram (**c**)

replacement of the affected valve, usually with cardiopulmonary bypass, and the concomitant complications and serious risks associated with open-heart surgery.

In the late 1990's, Professor Philipp Bonhoeffer developed a new technique of heart valve replacement that avoided the need for surgery (Bonhoeffer et al. 2000). This was based on the concept that a heart valve sewn inside a stent (Fig. 6a) could be reduced in size, by crimping it onto a balloon catheter (Fig. 6b), and then introduced through a peripheral vessel to the desired implantation site in the heart. Inflation of the balloon deploys the valved stent and anchors it within the old dysfunctional valve (Fig. 6c). Over the last 9 years, this simple technique has shown a marked learning curve with safety and efficacy improvements that have led to the successful, worldwide clinical use of this procedure in over 1,500 implants in the pulmonary position (Lurz et al. 2008) and thousands of implants in the aortic position (Chiam and Ruiz 2009).

Percutaneous pulmonary valve implantation (PPVI) potentially avoids multiple open-heart surgeries in patients with dysfunctional pulmonary valve both to reduce regurgitation and to treat stenosis. The device is made of a valve from a bovine jugular vein, sewn into a balloon-expandable, platinum-10% iridium stent (Fig. 6a). Bovine jugular venous valves are available only up to 22 mm of diameter, thus limiting the use of this minimally-invasive technique to those patients who have a small implantation site. Selection of patients for PPVI is critical to ensure procedural success and to guarantee long-term positive outcome, with no stent fracture occurrence (Khambadkone et al. 2005). Surgical history alone is not enough to decide whether a patient is suitable for the procedure or not; assessment of the implantation site with magnetic resonance (MR) imaging provides important additional information, but a complete understanding of the three-dimensional (3D) anatomy of the implantation site is necessary, not only to guarantee the success of this procedure, but also to study new devices in order to broaden the range of patients who could benefit from this treatment.

The aim of this work was to improve the understanding of the effects of patient specific implantation site morphologies on PPVI device mechanical performances, using a combination of imaging and engineering techniques.

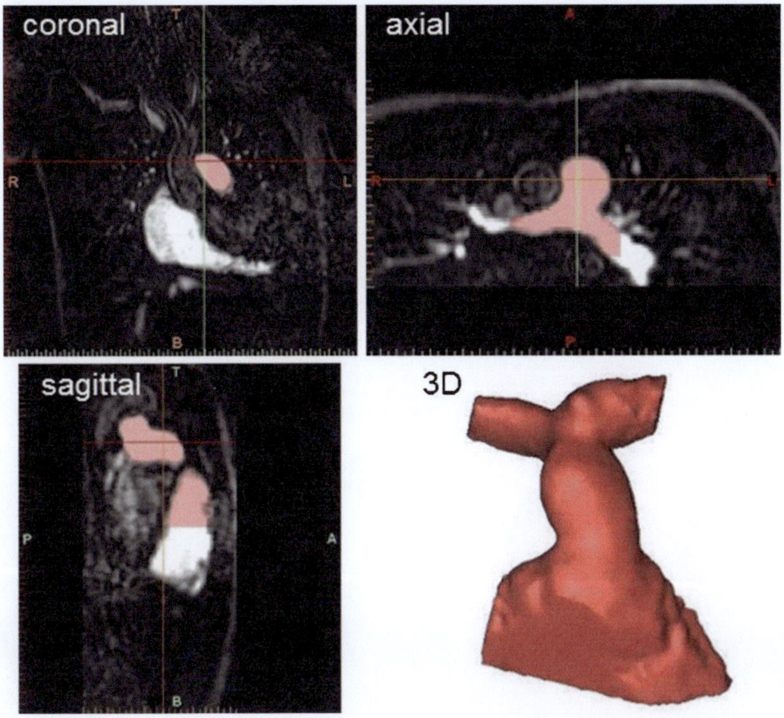

Fig. 7 Two-dimensional coronal, axial and sagittal views, and 3D reconstructed volume of a selected patient's RVOT

2.1 Materials and Methods

MR and computerized tomography (CT) images are the input data to model cardiac structures because, conveniently elaborated, they provide a reliable representation of the patient anatomy. The raw dicom data of patient MR/CT angiograms can be imported into commercial software (Materialise Inc., Leuven, Belgium) to reconstruct the implantation site anatomy of each individual patient. Segmentation masks detect the region of interest – the pulmonary outflow tract – in the coronal, axial and sagittal planes (Fig. 7). The software calculates a 3D model of the implantation site (blood volume = inner wall) for each potential PPVI candidate (Schievano et al. 2007a)

Starting from this 3D reconstruction, different modelling strategies were developed (Fig. 8). First, patients' implantation site morphology, size and dynamics were completely characterised in 3D. Second, physical objects of selected patients' implantation sites were created using rapid prototyping (RP) techniques. Third, the 3D volumes were used as input data for patient specific finite element (FE) analyses.

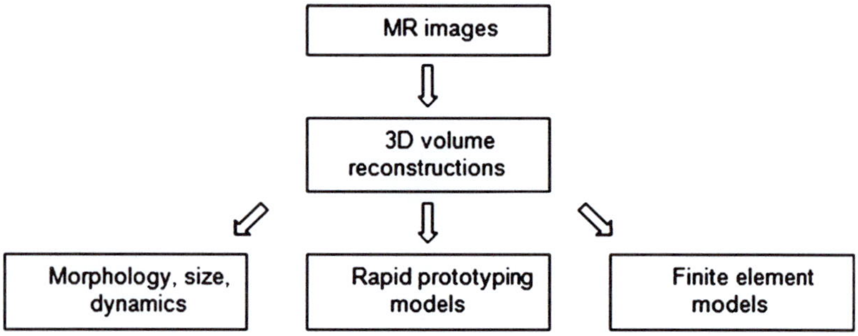

Fig. 8 Modelling cascade adopted in this study

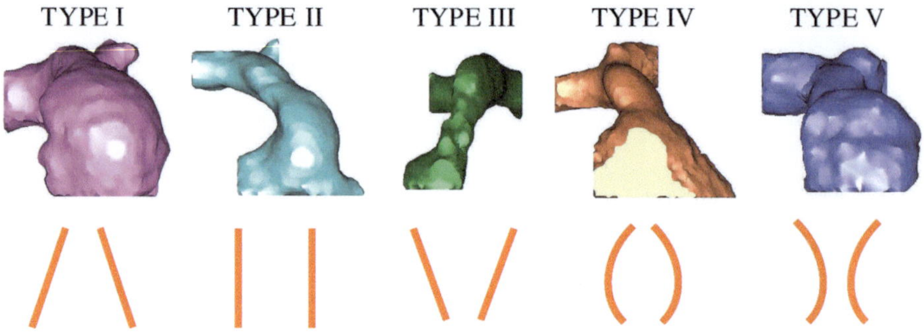

Fig. 9 Five types of RVOT morphology

2.1.1 Three-Dimensional Volume Analysis

We retrospectively studied 83 consecutive patients who were referred to our Institution for MR imaging assessment of right ventricular outflow tract (RVOT) dysfunction following previous repair of congenital heart disease (Schievano et al. 2007b). The primary diagnosis, RVOT type and number of open-heart surgeries were documented for this population. Ethical approval for MR imaging was obtained and all subjects (and/or a parent/guardian) gave informed consent for retrospective data analysis.

A morphological classification was created based on visualization of the patients' RVOT. Five types of RVOT anatomies were identified (Fig. 9): Type I had a pyramidal shape (wide proximally and narrow distally); Type II had a constant diameter; Type III had an inverted pyramidal shape (narrow proximally and wide distally); Type IV was wide centrally but narrowed proximally and distally; Type V was narrowed centrally but wide proximally and distally. In addition, the RVOT volumes were divided into sections and the shape of the cross-sectional areas standardized to circles. The diameter of these circles was calculated and graphically represented along the RVOT length. During PPVI, pressure inflation of the balloon deploying the device usually achieves a cylindrical

Fig. 10 The patient's RVOT wall geometry (**a**) is used as input file for a 3D printer (**b**), which adds and bonds material in layers, in order to fabricate the physical object (**c**)

configuration. It is therefore the diameter of this final circle which will influence the stability of the device and will guarantee the success of the procedure.

Furthermore, a four-dimensional (4D) CT image scan was performed in 12 patients to acquire 3D information over ten frames of the cardiac cycle (Schievano et al. 2011). These data were used to reconstruct 3D volumes of the implantation site and measure 3D deformation. The volumes were divided into sections and diameter changes over the cardiac cycle were measured at each section.

2.1.2 Rapid Prototyping

Rapid prototyping was used to fabricate physical models of 12 patients, who were selected to reflect the wide range of RVOT anatomies that are seen in our practice (Schievano et al. 2007a). Four of these patients underwent PPVI successfully; four went forward to catheterization but had an unsuccessful outcome – these patients represent those with borderline morphology, and the remaining four patients underwent surgery, as their anatomy was clearly too large for PPVI.

To establish the accuracy of the 3D RP models for selecting suitable patients for PPVI, we asked two expert interventional cardiologists, who were blinded to the outcome of the procedures, to re-assess the feasibility of the procedure in this group of patients, first on the basis of the conventional MR image alone; then, on the basis of the 3D RP models alone.

The standard stereolithography format of the patients' RVOT blood volume was exported to a RP machine (3D printer with a drop-on-demand technique (Armillotta et al. 2007, Fig. 10b) where a wall of 2 mm virtual thickness was built around the blood volume, to create the 3D model of the arterial wall (Fig. 10a).

2.1.3 Finite Element Analysis

Starting from the 3D reconstructed volume of selected patients' RVOTs, an FE model of the implantation site was created to virtually simulate the catheter procedure. An FE model of the current PPVI stent was also prepared (Schievano

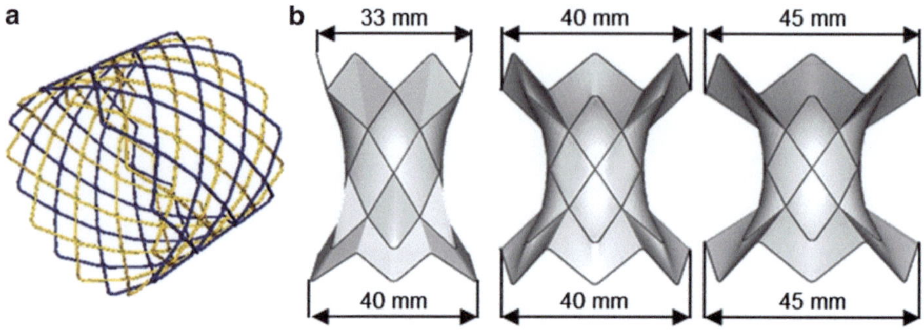

Fig. 11 Stent-in-stent design of the current PPVI device (**a**), and of a potential new PPVI device for lager outflow tract in three different dimensions (**b**)

et al. 2007c): the stent geometry was reconstructed from caliper and optic microscope measurements; the manufacturing company supplied material data. Contact algorithms were set up in the FE code (ABAQUS/Standard, Simulia, USA), and the PPVI stent model was inflated inside the RVOT model. The stresses induced by the stent on the arterial wall and the stress distribution on the device itself were analysed to verify safe anchoring and stent mechanical performance.

Using FE modelling, a 'stent-in-stent' technique (Fig. 11a) was investigated as a possible solution to common occurrence of stent fractures (Schievano et al. 2007c), which remains the major technical issue of the current PPVI device, occurring in approximately 20% of the patients (Nordmeyer et al. 2007).

To extend the indication of the percutaneous approach to the whole range of possible RVOT dimensions, innovative solutions that enable downsizing of the vessel diameter to the available valves, have been designed and tested, such as stent-graft geometries with a narrowing in the central portion that hosts the valve, and wider terminal parts that anchor the device to the RVOT wall (Fig. 11b) (Capelli et al. 2010). These stents are made of nitinol, a shape memory alloy with important biocompatibility, fracture resistance and MR compatibility properties and the peculiar characteristic of recovering very large deformation. In this study, patient-specific computational analyses have been applied to investigate the suitability of new device designs, using imaging data from 62 patients who had undergone surgical pulmonary valve replacement.

2.2 *Results*

2.2.1 Three-Dimensional Volume Analysis

RVOT morphology was heterogeneous; nevertheless 98% of all RVOT anatomies were described according to the defined classification (Fig. 12). Type I was the commonest morphology with 50% of the patients in this class. Thirteen percent of

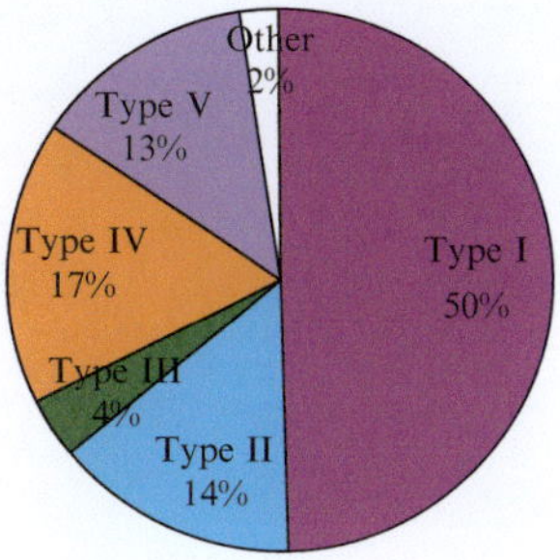

	PPVI	%
Type I	0/41	0
Type II	2/12	16
Type III	2/3	67
Type IV	3/14	21
Type V	3/11	27
Other	1/2	50
TOTAL	**11/83**	**13**

Fig. 12 Distribution of patients in the five morphological classes and number of PPVIs performed in the studied population

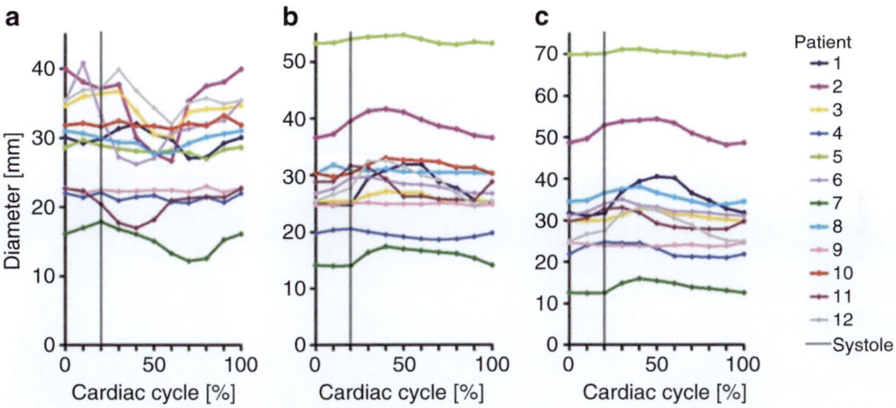

Fig. 13 RVOT diameter changes from 4DCT in three different sections along the pulmonary outflow tract (A = sub-valve, B = supra-valve, C = pre-bifurcation) of 12 patients

patients underwent PPVI (Fig. 12): none of them had Type I morphology, which was felt unsuitable because of the high risk of proximal device dislodgement.

From a morphological perspective, we could have attempted PPVI in all types of RVOT except Type I. In fact, we performed PPVI only in 11 out of 42 patients with suitable morphology, as other factors influence patient selection for this procedure, such as the dimension of the implantation site, which cannot be greater than 22 mm. Diameter measurements clearly show that most of the analysed RVOTs were larger than 22 mm. In addition, suitable implantation sites must be rigid conduits which can guarantee protection from fracture, while the dynamics that developed in these patients late after surgical repair of congenital heart disease presented large deformations in all directions, as shown by the 4DCT analysis (Fig. 13).

The development of larger devices with a new design that permits downsizing of dilated RVOTs to the size of biologically available valves will significantly increase the number of patients that could benefit from a percutaneous approach in the future.

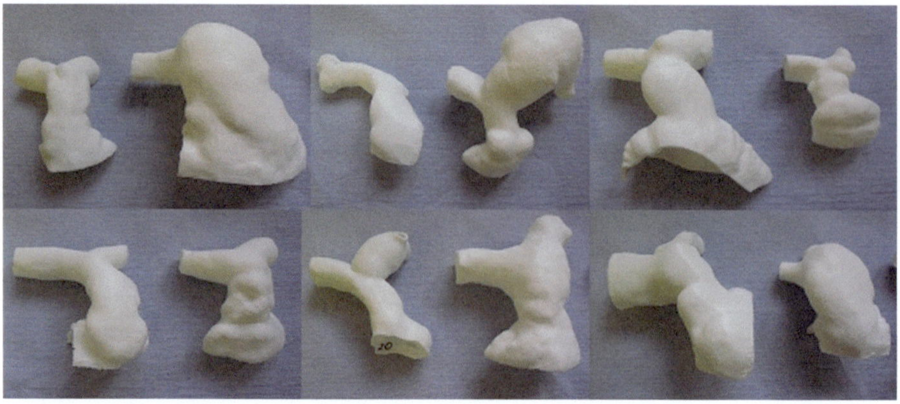

Fig. 14 3D RP models of 12 patients' RVOT

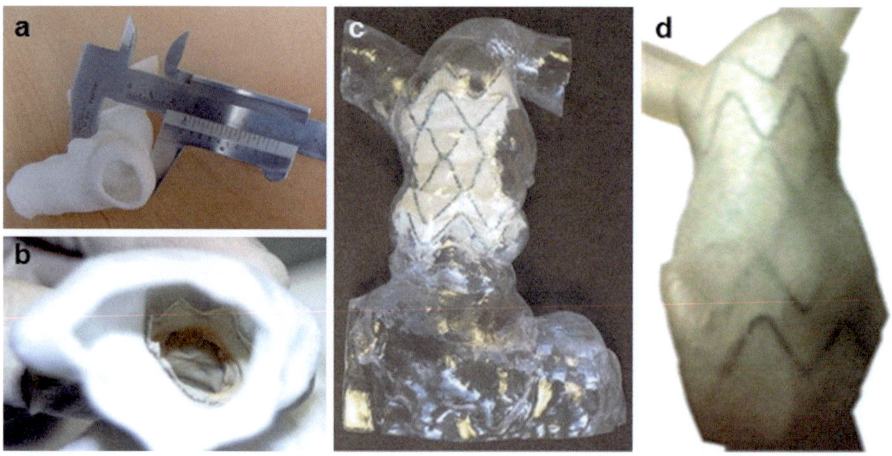

Fig. 15 Calliper measurements (**a**) and *in-vitro* trial insertion of the current PPVI device (**b**) and of a new prototype (**c**, **d**) in different patients' RP models

2.2.2 Rapid Prototyping

The RP results are shown in Fig. 14. The cardiologists made the correct decision in 50% and 66% of the cases respectively, using conventional MR assessment, which improved for both to 75% with the availability of the 3D RP models. In the patients where PPVI was attempted but failed, the decisions were 50% correct with the MR images alone, but 100% correct using the models for both observers.

Physical RP models can help the selection of patients for PPVI more accurately than conventional MR because they provide a complete appreciation of the 3D RVOT anatomy and enable easy measurement of RVOT size (Fig. 15a). The models also enabled trial implantation of the PPVI device (Fig. 15b) to test the correct

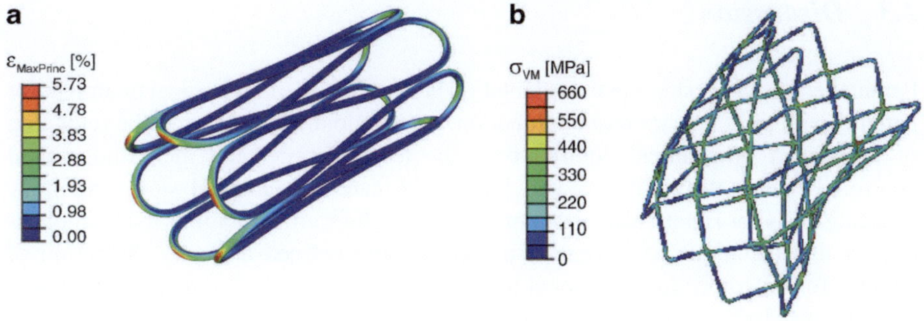

Fig. 16 Von Mises stress distribution in the current PPVI stent after FE inflation in a patient's implantation site (**a**) and an example of an optimised, nitinol ring for a new PPVI stent (**b**) – maximum principal strain contour when the ring is crimped inside the catheter

Fig. 17 Implantation of a new stent graft in selected patients' outflow tracts

positioning and anchoring into the RVOT before the procedure was actually performed for borderline patients. This guaranteed a more accurate selection process for PPVI and enhanced patient safety. Furthermore, the RP models were used to test prototypes of new possible devices for PPVI (Fig. 15c, d) (Schievano et al. 2010).

2.2.3 Finite Element Analysis

The FE analysis of the current stent identified the areas at potential risk for fracture as seen in clinical practice (Fig. 16a). The stent-in-stent technique appeared to be an effective solution to increase the device mechanical performance and reduce the rate of fracture. However, the design of this stent would still not be suitable to treat patients who have large RVOTs. Purposely designed nitinol stent grafts could be the optimal choice for the next generation PPVI device. FE analysis can aid the optimisation of these devices (Fig. 16b) before prototypes are manufactured and tested in RP models.

FE analyses allowed the evaluation of the stresses not only in the stent but also those induced in the RVOT wall by the deployment of the device. These stresses should be not too elevated to avoid wall damage, but also high enough to hold the stent in the correct position, thus avoiding valve dislodgment.

Deployment and anchoring of a potential new PPVI stent-graft with three different dimensions were tested in patient specific implantation sites using FE analyses (Fig. 17). By varying the dimensions of the device, the number of patients that could potentially benefit from PPVI can be increased to 70% (Capelli et al. 2010).

2.3 Discussion

Percutaneous pulmonary valve implantation is a successful, non-surgical treatment for patients with pulmonary valve dysfunction, with undoubted benefits in the reduction of open-heart procedures required over the total lifetime, and increased patient comfort because it avoids prolonged intensive care and hospital stay, ultimately reducing costs for Health Care systems. However, there are limitations to the application of this technique to patients with specifically defined anatomy. Selection of patients for this procedure is critical to ensure procedural success, patient safety and to guarantee long-term positive outcome. This study describes the application of imaging and engineering methodologies to improve the understanding of the mechanisms of PPVI, in order to enhance the results of this minimally-invasive technique and increase the number of patients who could benefit from this procedure.

Using 3D reconstructions of patient specific implantation site morphologies from MR images, the study of the implantation site morphology identified subgroups of RVOT morphology that would be suitable for PPVI and documented their prevalence in our patient population. Morphology of the RVOT together with size and dynamics are the major determinants of suitability for PPVI, highlighting the importance of 3D imaging reconstructions and measurements in patient assessment. The findings of this study identified the need for a valved stent that may be able to deal with this wide range of morphology, size and dynamics. The morphological classification developed during this research can guide the anatomical and dimensional requirements that the next generation device will need to fulfil.

Establishing the precise morphology of the proposed implantation site with the currently available techniques (echocardiography, MR, CT imaging) is extremely valuable, but we have demonstrated that in borderline cases it can still be difficult to perceive 3D anatomy that is presented on a 2D computer screen, and we have had several cases in which the transcatheter procedure has been attempted, but has not been successful. Starting from the RVOT 3D volume reconstruction, RP models can be used by the clinicians in the decision process regarding the suitability for transcatheter procedure, because they provide a complete appreciation of the 3D anatomy and size of the implantation site and enable trial implantation of the device. The RP models of some patients' RVOTs, selected among the ones which represent the currently untreatable morphologies, can also be used as bench test for new devices.

Finite element analysis allowed the mechanical performance of the current PPVI stent to be understood, in particular with regard to a common complication of the device – stent fracture – which can lead to major clinical events. This led to the development of the 'stent-in-stent' concept, which appears to be an effective solution to reduce the rate of device fracture and increase the success of the percutaneous procedure, without compromising its technical ease. The design of the multi-element stent, however, is still not suitable to treat the numerous patients who have large RVOTs. A first attempt towards next generation PPVI stents, with innovative designs that can broaden the applications of PPVI, was reported in this study. Finite element modelling was used to test these new devices in patient specific implantation site, and guide the optimisation process towards a robust final design.

3 Conclusions

Engineering methodologies, together with advanced imaging technologies, can be used to replicate patient specific patho-physiological conditions in a virtual model. Combining patient specific imaging data and computational modelling offers a new tool to obtain additional, predictive information about responses to therapy in individual patients. Numerical techniques can therefore improve our understanding of the heart structures and the way devices interact with them, ultimately aiming at optimising the treatment for patients and progressing in the field of cardiac applications.

References

Armillotta A, Bonhoeffer P, Dubini G, Ferragina S, Migliavacca F, Sala G, Schievano S (2007) Use of rapid prototyping models in planning of percutaneous pulmonary valved stent implantation. Proc Inst Mech Eng H 221(4):407–416

Blackstone E (2005) Could it happen again? The Björk-Shiley convexo-concave heart valve story. Circulation 111:2717–2719

Bonhoeffer P, Boudjemline Y, Saliba Z, Merckx J, Aggoun Y, Bonnet D et al (2000) Percutaneous replacement of pulmonary valve in a right-ventricle to pulmonary-artery prosthetic conduit with valve dysfunction. Lancet 9239:1403–1405

Capelli C, Taylor AM, Migliavacca F, Bonhoeffer P, Schievano S (2010) Patient-specific reconstructed anatomies and computer simulations are fundamental for selecting medical device treatment: application to a new percutaneous pulmonary valve. Phil Trans R Soc A 368:3027–3038

Chandran KB, Dexter EU, Aluri S, Richenbacher W (1998) Negative pressure transients with mechanical heart-valve closure: correlation between in vitro and in vivo results. Ann Biomed Eng 26:4

Chiam PT, Ruiz CE (2009) Percutaneous transcatheter aortic valve implantation: evolution of the technology. Am Heart J 157(2):229–242

Díaz-Zuccarini V, LeFèvre J (2007) An energetically coherent lumped parameter model of the left ventricle specially developed for educational purposes. Comput Biol Med 37:774–784

Díaz-Zuccarini V, Rafiroiu D, Hose DR, Lawford PV, Narracott AJ (2008) Multiphysics and multiscale simulation: application to a coupled model of the left ventricle and a mechanical heart valve. Int J Multiscale Comput Eng 6:65–76

Díaz-Zuccarini V, Rafiroiu D, LeFevre J, Hose DR, Lawford PV (2009) Systemic modelling and computational physiology: the application of Bond Graph boundary conditions for 3D cardiovascular models. Simul Model Pract Theory 17:125–136

Fielder J (1993) Getting the bad news about your artificial heart valve. Hastings Cent Rep 23 (2):22–28

Formaggia L, Gerbeau JF, Nobile F, Quarteroni A (2001) On the coupling of 3D and 1D Navier-Stokes equations for flow problems in compliant vessels. Comput Methods Appl Mech Eng 191:561–582

Hose DR, Narracott AJ, Penrose J, Baguley D, Jones IP, Lawford PV (2006) Fundamental mechanics of aortic valve closure. J Biomech 39:958–967

Jones DM, Fenner JW, Berti G (2004) The GEMSS grid: an evolving HPC environment for medical applications. In: Proceedings of HealthGrid, Clermont Ferrand

Khambadkone S, Coats L, Taylor A, Boudjemline Y, Derrick G, Tsang V et al (2005) Percutaneous pulmonary valve implantation in humans: results in 59 consecutive patients. Circulation 112(8):1189–1197

Lurz P, Coats L, Khambadkone S, Nordmeyer J, Boudjemline Y, Schievano S, Muthurangu V, Lee TY, Parenzan G, Derrick G, Cullen S, Walker F, Tsang V, Deanfield J, Taylor AM, Bonhoeffer P (2008) Percutaneous pulmonary valve implantation – impact of evolving technology and learning curve on clinical outcome. Circulation 117(15):1964–1972

Migliavacca F, Balossino R, Pennati G, Dubini G, Hsia T, de Leval M, Bove E (2005) Multiscale modelling in biofluidynamics: application to reconstructive paediatric cardiac surgery. J Biomech 39:1010–1020

Nordmeyer J, Khambadkone S, Coats L, Schievano S, Lurz P, Parenzan G, Taylor AM, Lock JE, Bonhoeffer P (2007) Risk stratification, systematic classification and anticipatory management strategies for stent fracture after percutaneous pulmonary valve implantation. Circulation 115 (11):1392–1397

Prendergast P, Maher SA (2001) Issues in pre-clinical testing of implants. J Mater Process Technol 118:337–342

Quarteroni A, Veneziani A (2003) Analysis of a geometrical multiscale model based on the coupling of PDE's and ODE's for blood flow simulations multiscale modeling and simulation. SIAM 1:173–195

Rafiroiu D, Díaz-Zuccarini V, Narracott A, Lawford PV, Hose DR (2008) Multi-physics and multi-scale modelling: closure dynamics of a bileaflet prosthetic heart valve. In: Proceedings of the European Society of Biomechanics, Lucerne, pp S241

Schievano S, Migliavacca F, Coats L, Khambadkone S, Carminati M, Wilson N, Deanfield J, Bonhoeffer P, Taylor A (2007a) Rapid prototyping three-dimensional model building of the right ventricular outflow tract and pulmonary trunk from magnetic resonance data – clinical utility for percutaneous pulmonary valve implantation. Radiology 242(2):490–497

Schievano S, Coats L, Migliavacca F, Norman W, Frigiola A, Deanfield J, Bonhoeffer P, Taylor AM (2007b) Variations in right ventricular outflow tract morphology following repair of congenital heart disease – implications for percutaneous pulmonary valve implantation. J Cardiovasc Magn Reson 9(4):687–695

Schievano S, Petrini L, Migliavacca F, Coats L, Nordmeyer J, Lurz P, Khambadkone S, Taylor AM, Dubini G, Bonhoeffer P (2007c) Finite element analysis of stent deployment: understanding stent fracture in percutaneous pulmonary valve implantation. J Interv Cardiol 20 (6):546–554

Schievano S, Taylor AM, Capelli C, Coats L, Walker F, Lurz P, Nordmeyer J, Wright S, Khambadkone S, Tsang V, Carminati M, Bonhoeffer P (2010) A new approach to medical device development – first-in-man implantation of a novel percutaneous valve. EuroIntervention 5:745–750

Schievano S, Capelli C, Lurz P, Nordmeyer J, Young C, Owens C, Bonhoeffer P, Taylor AM (2011) Four-dimensional computed tomography: a method of assessing right ventricular outflow tract and pulmonary artery deformations throughout the cardiac cycle. Eur Radiol 21 (1):36–45

Westerhof N, Bosman F, De Vries CJ (1969) Analog studies of the human systemic arterial tree. J Biomech 2:121–143

Yoganathan AP, He Z, Jones C (2004) Fluid mechanics of heart valves. Annual review of biomedical engineering, American Heart Association. Heart Dis Stroke Stat 6:331–362

Numerical Simulation of Viscous Flow Around Kayak: A Comparison of Different Design Models

Daniel A. Marinho, Vishveshwar R. Mantha, Abel I. Rouboa, and António J. Silva

Abstract The aim of this paper was to analyze the viscous flow around different kayak design models, allowing computing hydrodynamic drag force through numerical simulations. The simulations were based on Finite volume method of discretization. Numerical simulations were performed on three different kayak models, corresponding to kayak design evolution of K1 Vanquish M Nelo™ models (M.A.R. Kayaks Lda, Portugal). The numerical simulations were performed only for the outer shell section of the kayak hull geometry, assumed to be submerged in still water. For a speed of 5.0 m/s, it can be observed that the pressure of the bow is larger, and the pressure of the stern is smaller, and variation of pressure of the middle is relatively small. The main results suggest that the evolution from Vanquish I to Vanquish III was succeeded, as shown by the reduction of hydrodynamic drag over the three models studied. In the design of Kayak, one can optimize the shape of the outer surface by combining the pressure and shear stress distribution of the shell.

Keywords Computational fluid dynamics • Sports • Water • Drag

D.A. Marinho (✉)
Departamento de Ciências do Desporto, Universidade da Beira Interior,
Rua Marquês de Ávila e Bolama, 6201-001 Covilhã, Portugal

Research Centre in Sports, Health and Human Development (CIDESD), Vila Real, Portugal
e-mail: dmarinho@ubi.pt

V.R. Mantha • A.J. Silva
Research Centre in Sports, Health and Human Development (CIDESD), Vila Real, Portugal

Department of Sport Sciences, Exercise and Health, University of Trás-os-Montes and Alto Douro (UTAD), Vila Real, Portugal

A.I. Rouboa
Research Centre in Sports, Health and Human Development (CIDESD), Vila Real, Portugal

Department of Engineering, University of Trás-os-Montes and Alto Douro (UTAD),
Vila Real, Portugal

D. Iacoviello and U. Andreaus (eds.), *Biomedical Imaging and Computational Modeling in Biomechanics*, Lecture Notes in Computational Vision and Biomechanics 4,
DOI 10.1007/978-94-007-4270-3_10, © Springer Science+Business Media Dordrecht 2013

1 Introduction

It is a generally held principle of boat hydrodynamics that the speed of a boat will be a function of the amount of power delivered, and the amount of resistance created by the water as the hull of the boat passes through it. It follows that the longer and slimmer the hull of a craft, the generally less pronounced the effect of friction will be. As the weight of the canoe is spread over a greater hull length, the hull will draw less water, and thus the craft will generally be less stable. Hydrodynamic research (Jackson 1995) regarding the most efficient canoe hull designs suggests that 90% of the drag on the boat is the water, while the remaining 10% is created by the hull and the paddlers moving through the air above the water.

The broad physical modelling capabilities of computer numerical simulations have been applied to industrial scope in a wide range of applications. Nowadays, several companies throughout the world benefit from this important engineering design and analysis tool. Its extensive range of multiphysics capabilities makes it an important and interesting tool in engineering studies. In sports scope, the main results suggested that numerical simulations could provide useful information about performance. Indeed, this methodology has produced significant improvements in equipment design and technique prescription in areas such as sailing performance (Pallis et al. 2000), Formula 1 racing (Kellar et al. 1999), winter sports (Dabnichki and Avital 2006), and swimming (Marinho et al. 2009).

Therefore, the aim of this paper was to analyze the viscous flow around different kayak design models, allowing computing hydrodynamic drag force through numerical simulations. Computational fluid dynamics methodology allows analyzing the fluid flow in each point of the kayak model. This reveals an important task to optimize the shape of the kayak, permitting also to verify differences between each kayak model. The authors attempted to study the evolution of one specific kayak model (K1 Vanquish M Nelo™, M.A.R. Kayaks Lda, Portugal) during three phases of this development process (Vanquish I, Vanquish II, Vanquish III).

This chapter covers topics in kayak simulation from a computational fluid dynamics perspective. This perspective means emphasis on the fluid mechanics and computational fluid dynamics methodology applied in the design of different kayak models. We concentrated on numerical simulation results, considering the scientific simulation point-of-view and especially the practical implications to enhance the development of the models.

2 Materials and Methods

2.1 Geometry and Computational Domain

Numerical simulations were performed on three different kayak models, corresponding to kayak design evolution of K1 Vanquish M Nelo™ models (M.A.R. Kayaks Lda, Portugal). Figure 1 presents one of the kayaks (K1 Vanquish IIIM) analysed in the current research.

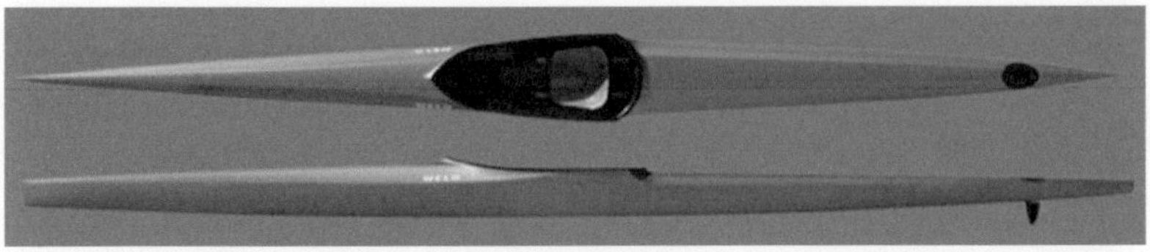

Fig. 1 K1 Vanquish IIIM model of kayak shown top view (*top*) and side view (*bottom*)

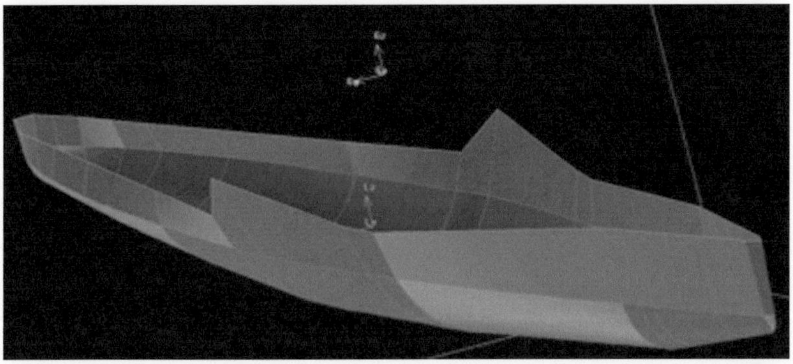

Fig. 2 The development of the geometry of K1 Vanquish IIIM model in Gambit software

To obtain the geometry of each kayak, models were developed in Gambit, a geometry-modelling program of Ansys Fluent^{TM} 6.3 commercial software (Ansys, Canonsburg, Pennsylvania, U.S.A.), which provides sophisticated computational fluid dynamics software (Fig. 2). The simulations were based on Finite volume method of discretization. These surfaces were then meshed, using the same program, creating the mesh, which has been imported into Fluent computational fluid dynamics program for analysis (Fig. 3).

The numerical simulations were performed only for the outer shell section of the kayak hull geometry, assumed to be submerged in still water.

2.2 Governing Equations

For a constant speed of a shell, the shell can be considered as static and the fluid can be considered as flow with some speed in opposite direction. The fluctuation of density of water was ignored, and account for the variation of average density. It was employed the average approach to the Navier–Stokes (RANS) equations, added the Reynolds stress to the turbulence models, and then a group equations of viscous flow around shell, that is RANS equations, could be obtained. The Navier–Stokes equations describe the motion of viscous non-compressible fluid substances. These equations arise from applying Newton's second law to fluid motion, together with the

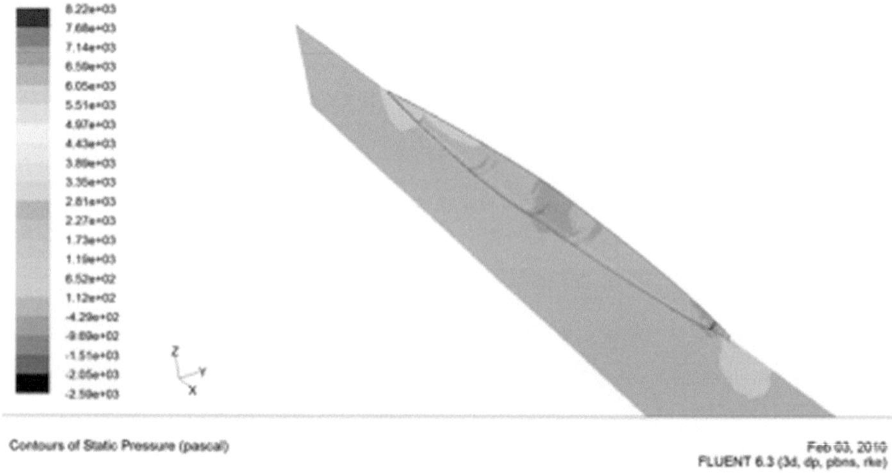

Fig. 3 K1 Vanquish IIIM model in Fluent software

assumption that the fluid stress is the sum of a diffusing viscous term (proportional to the gradient of velocity), plus a pressure term. A solution of the Navier–Stokes equations is called a velocity field or flow field, which is a description of the velocity of the fluid at a given point in space and time.

The flow around the kayak is turbulent. We have considered Reynolds averaged Navier–Stokes equations. The closure problem of the turbulent modeling is solved using k-epsilon model with appropriate wall functions (Launder and Spalding 1974). The system of equations for solving 3D, incompressible fluid flow in steady-state regime and the averaged-RANS equations are as follows:

Continuity equation:

$$\frac{\partial y}{\partial x_i}(\bar{U}_i) = 0 \tag{1}$$

where, i = 1, 2, 3

Navier–Stokes (momentum) equations:

$$\frac{\partial y}{\partial x_j}(\rho\bar{U}_i\bar{U}_j) = -\frac{\partial \bar{p}}{\partial x_i} + \frac{\partial}{\partial x_j}\left[(\mu + \mu_t)\left(\frac{\partial}{\partial x_j}(\bar{U}_i) + \frac{\partial}{\partial x_i}(\bar{U}_j)\right) - \frac{2}{3}\delta_{ij}\rho k\right] \tag{2}$$

The equations of turbulence model:

$$\frac{\partial k}{\partial x_i}(\rho\bar{U}_i) = \frac{\partial}{\partial x_j}\left[\left(\mu + \frac{\mu_t}{\sigma_k}\right)\frac{\partial k}{\partial x_j}\right] + P_k - \rho\varepsilon \tag{3}$$

$$\frac{\partial \varepsilon}{\partial x_i}(\rho\bar{U}_i) = \frac{\partial}{\partial x_j}\left[\left(\mu + \frac{\mu_t}{\sigma_\varepsilon}\right)\frac{\partial \varepsilon}{\partial x_j}\right] + C_1 P_k\frac{\varepsilon}{k} - C_2\rho\frac{\varepsilon^2}{k} \tag{4}$$

Table 1 Modified k-ε model constants

C_μ	C1	C2	σ_k	σ_ε
0.09	1.44	1.92	1.0	1.3

The production of kinetic energy term is given as:

$$P_k = -\rho\overline{u'_i u'_j}\frac{\partial \bar{U}_j}{\partial x_i} = \mu_t S^2 \tag{5}$$

where, S is Strain rate given by

$$S \equiv \sqrt{2S_{ij}S_{ij}} \tag{6}$$

The turbulent viscosity is calculated by the following relation:

$$\mu_t \equiv \rho C_\mu \frac{k^2}{\varepsilon} \tag{7}$$

where, $U_i(t) \equiv \bar{U}_i + u'_i$ is the component of instantaneous velocity in i-direction (m/s), $\bar{U}_i$ is the component of time averaged mean velocity in i-direction (m/s), u'_i is the component of fluctuating velocity in i-direction (m/s), i, j are the direction vectors, ρ is average fluid density (kg/m^3), μ is dynamic viscosity of fluid (kg/ms), μ_t is turbulent viscosity of fluid (kg/ms), $\bar{p}$ is average pressure, $k = \bar{u}_i\bar{u}_j/2$ is the turbulent kinetic energy per unit mass (m^2/s^2), ε is a dissipation rate of the turbulent kinetic energy per mass unit (m^2/s^3), δ_{ij} is the Kronecker delta with the condition that: $\delta_{ij} = 1$ if $i = j$ and $\delta_{ij} = 0$ if $i \neq j$.

The turbulent model constants σ_k, σ_ε, C_1, C_2 are shown in Table 1.

2.3 Boundary Condition

The boundary conditions imposed in this study are as follows: At the inlet, the average uniform velocity equals to 5 m/s, corresponding to Froude number of order of 7. The inlet is provided with turbulent intensity of 1%. At the outlet, the relative pressure of the fluid is null (outflow). The upper and lower, front and back domain limits are provided with symmetry condition (Fig. 4).

2.4 Numerical Scheme

The simulations are based on Finite volume method of discretization (Patankar 1980). In order to limit numerical dissipation, particularly when the geometry is complex inducing an unstructured grid, second order discretization schemes are

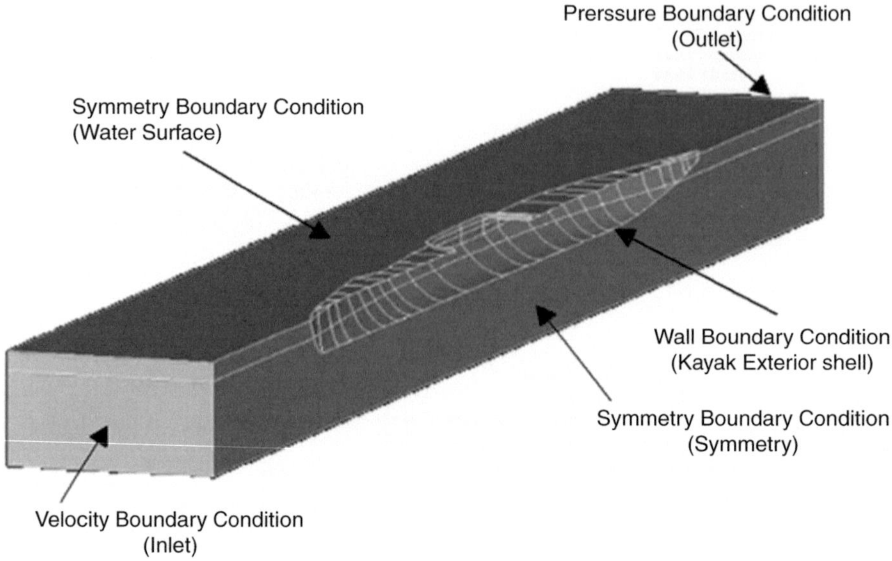

Fig. 4 The schematic of K1 Vanquish IIIM model showing the applied boundary conditions in Fluent software

used. Semi-Implicit Method for Pressure Linked Equations (SIMPLE) algorithm is used for the solution process. The discretized non-linear algebraic equations are solved by Tri-Diagonal Method Algorithm (TDMA) iterative method. The tetrahedral grid with 10,130,418 cells was used for computation. In generic terms, the convergence of the calculation is checked by the value of the residuals of the various flow parameters. Default values in Ansys Fluent are set at 10^{-3}. This criterion is assumed sufficient to ensure the convergence of the solution for the present study (Bixler et al. 2007).

3 Results and Discussion

Computational Fluid Dynamics methodology allows analyzing the fluid flow in each point of the kayak model. This reveals an important task to optimize the shape of the kayak, permitting also to verify differences between each kayak model (Vanquish I, Vanquish II, Vanquish III).

Following figures present velocity distribution along the middle line of longitude of the kayak (Fig. 5), viscous coefficient distribution on surface of the kayak (Fig. 6), pressure coefficient distribution on surface of the kayak (Fig. 7), and velocity vector field (Fig. 8). It can be observed that the pressure of the bow is larger, and the pressure of the stern is smaller, and variation of pressure of the

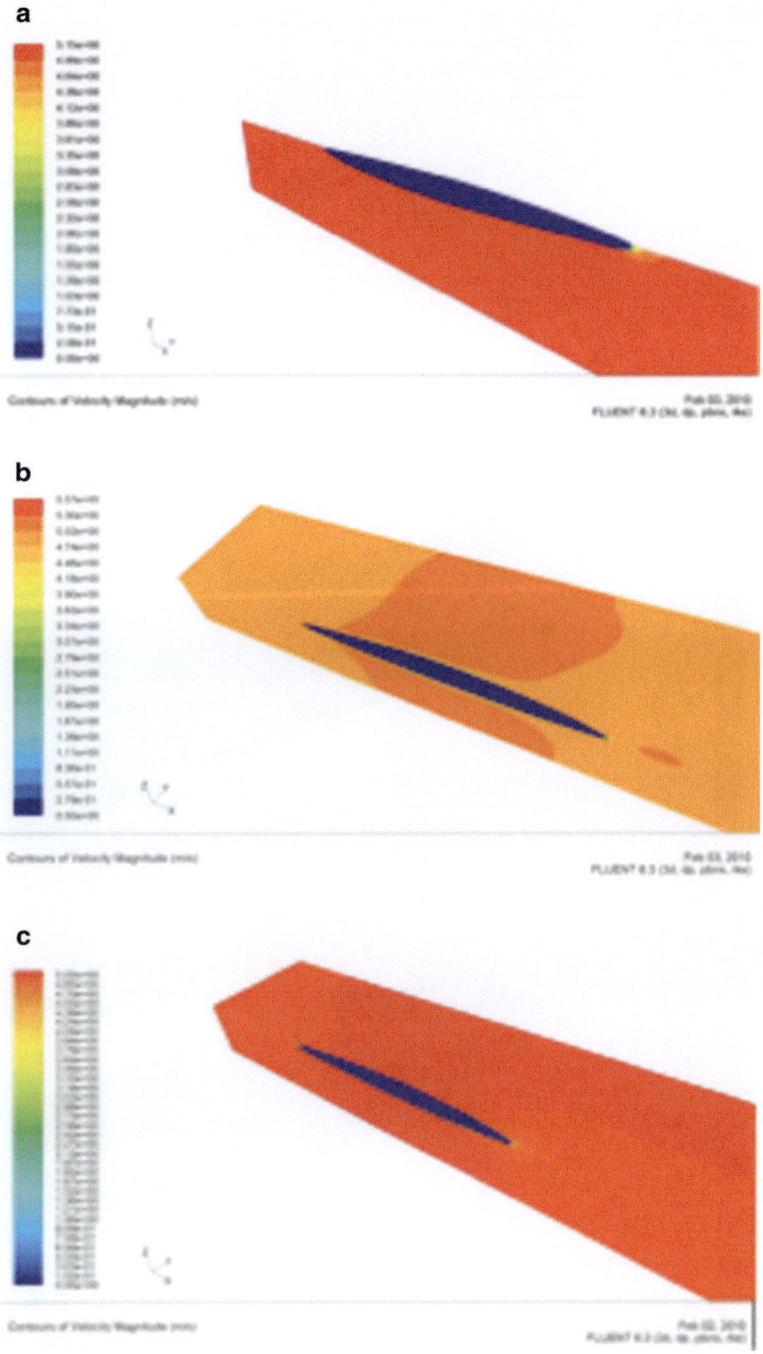

Fig. 5 K1 Vanquish M models ((**a**) Vanquish I, (**b**) Vanquish II, (**c**) Vanquish III) of kayak showing contours of velocity magnitude

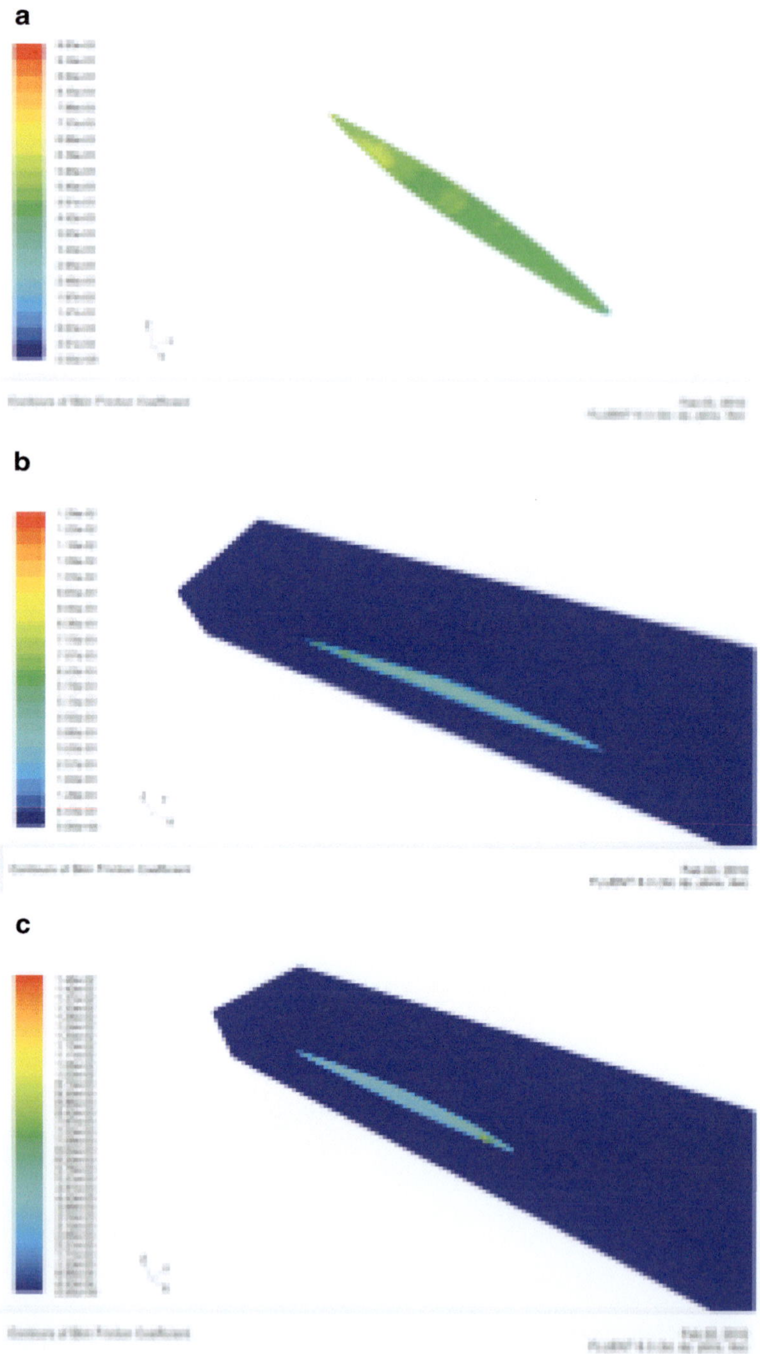

Fig. 6 K1 Vanquish M models ((**a**) Vanquish I, (**b**) Vanquish II, (**c**) Vanquish III) of kayak showing contours of viscous coefficient

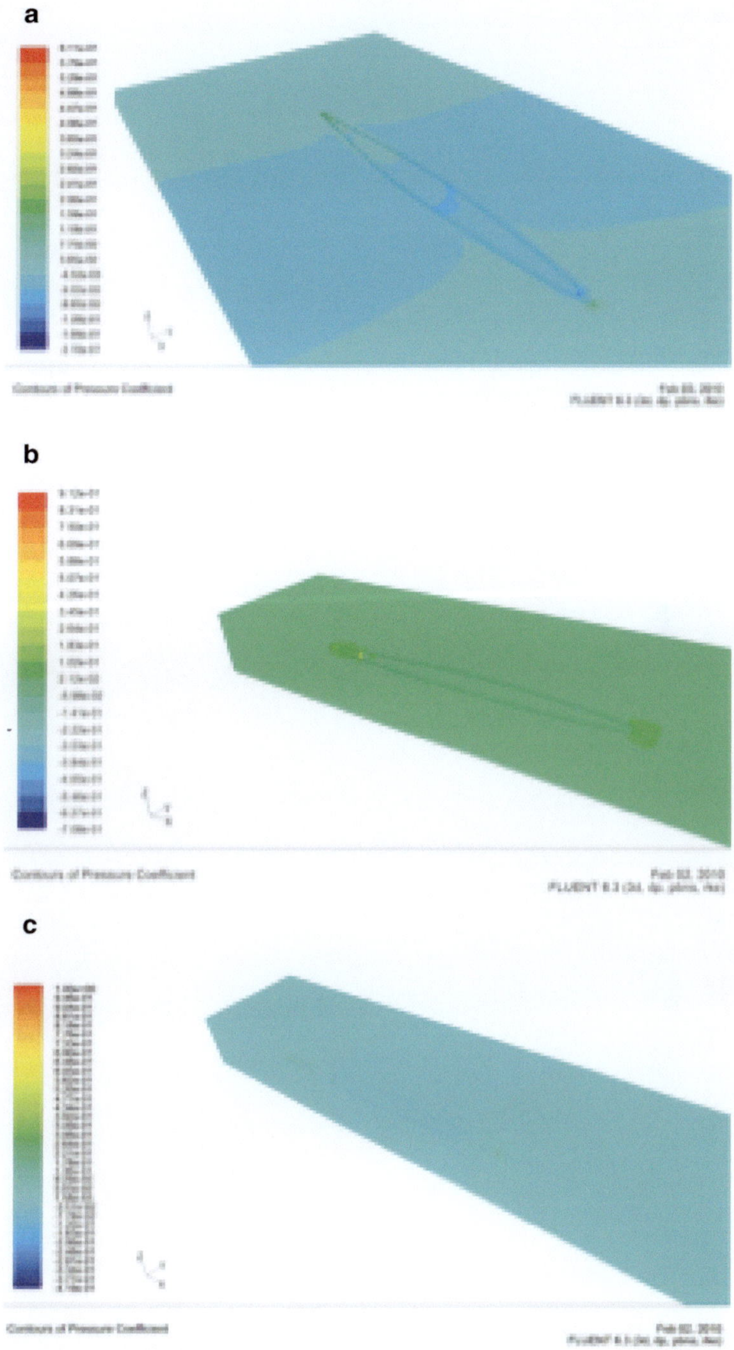

Fig. 7 K1 Vanquish M models ((**a**) Vanquish I, (**b**) Vanquish II, (**c**) Vanquish III) of kayak showing contours of pressure coefficient

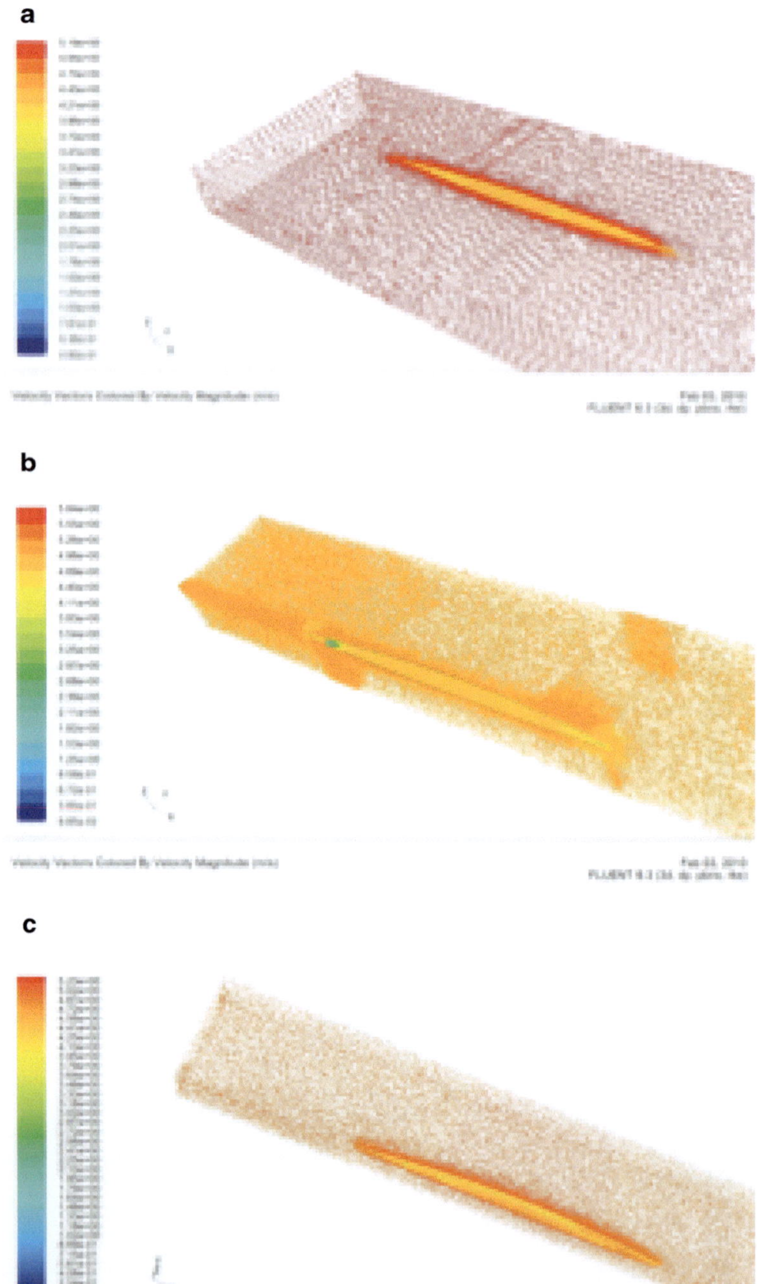

Fig. 8 K1 Vanquish M models ((**a**): Vanquish I, (**b**): Vanquish II, (**c**): Vanquish III) of kayak showing velocity vector field

Table 2 The comparative results of pressure drag coefficient and viscous drag coefficient for three kayak models

Kayak model	Velocity	Pressure drag coefficient	Viscous drag coefficient
K1 Vanquish IM	5.0	3.92e-3	11.44e-3
K1 Vanquish IIM	5.0	1.39e-3	3.65e-3
K1 Vanquish IIIM	5.0	0.59e-3	3.67e-3

middle is relatively small. Additionally, some differences can be noticed between different models of kayaks, especially when one adds data from Table 2.

Table 2 shows the results of pressure drag coefficient and viscous drag coefficient obtained for the three models of the kayaks. It is important to underline the important reduction on both drag parameters from K1 Vanquish IM to K1 Vanquish IIM. From K1 Vanquish IIM to K1 Vanquish IIIM, there is also an important reduction on pressure drag, although viscous drag remained almost constant.

In the design of kayak, one can optimize the shape of the outer surface by combining the pressure and shear stress distribution of the shell according to the above results.

4 Conclusions

The main results suggest that the evolution from Vanquish I to Vanquish III was succeeded. Pressure drag decreased significantly from Vanquish I to Vanquish II and again to Vanquish III, presenting a very small value in this last model. This fact suggests that the shape of the boat presents a very good hydrodynamic profile and the kayak evolution is in a good direction.

Regarding viscous drag, a significant decrease was obtained from Vanquish I to Vanquish II, although the values remained almost constant from Vanquish II to Vanquish III. The intensity of the values could suggest that viscous drag could be an important parameter to be analyzed in further studies and to be "performed" in future boat developments. In fact, pressure drag was reduced a lot, and viscous drag represents the large component of drag at this stage. The decrease of boundary layer throughout, for instance, roughness decrease and/or "allowing" sliding conditions at the surface of the boat could be a good solution. However, this approach should be experimentally tested before fully applied. Moreover, in this study, only 2-D models were tested and only the submerged part of the boat was considered in the simulations. Thus, more accurate results could be achieved if whole 3-D models of the kayaks were tested, considering also the part of the boat above the water line, situation that could be carried-out continuing this important cooperation with the company.

Acknowledgments Our research has been supported by the Portuguese Government and European Union by grants of the Science and Technology Foundation (PTDC/DES/098532/2008; FCOMP-01-0124-FEDER-009569) and QREN (Process no 4608, SI I&DT 17/2008, Vale I&DT).

The authors would also like to acknowledge Manuel Ramos (Nelo), Nuno André Santos, João Passos, and João Costa, from MAR KAYAKS, LDA Company for all the support during this project.

References

Bixler B, Pease D, Fairhurst F (2007) The accuracy of computational fluid dynamics analysis of the passive drag of a male swimmer. Sports Biomech 6:81–98

Dabnichki P, Avital E (2006) Influence of the position of crew members on aerodynamics performance of two-man bobsleigh. J Biomech 39:2733–2742

Jackson PS (1995) Performance prediction for Olympic Kayaks. J Sports Sci 13:239–245

Kellar WP, Pearse SRG, Savill AM (1999) Formula 1 car wheel aerodynamics. Sports Eng 2:203–212

Launder BE, Spalding DB (1974) The numerical computation of turbulent lows. Int J Comput Meth Appl Mech Eng 3:269–289

Marinho DA, Reis VM, Alves FB, Vilas-Boas JP, Machado L, Silva AJ, Rouboa AI (2009) The hydrodynamic drag during gliding in swimming. J Appl Biomech 25:253–257

Pallis JM, Banks DW, Okamoto KK (2000) 3D computational fluid dynamics in competitive sail, yatch and windsurfer design. In: Subic F, Haake M (eds) The engineering of sport: re-search, development and innovation. Blackwell Science, Oxford, pp 75–79

Patankar SV (1980) Numerical heat transfer and fluid flow. Hemisphere Publishing Corporation, Washington DC

MIX
Papier aus verantwortungsvollen Quellen
Paper from responsible sources
FSC® C105338

If you have any concerns about our products,
you can contact us on
ProductSafety@springernature.com

In case Publisher is established outside the EU,
the EU authorized representative is:
Springer Nature Customer Service Center GmbH
Europaplatz 3, 69115 Heidelberg, Germany

Printed by Libri Plureos GmbH
in Hamburg, Germany